RECONSTRUCTING THE SCIENCE OF HEAT FOR BETTER TEACHING AND LEARNING

RECONSTRUCTING THE SCIENCE OF HEAT FOR BETTER TEACHING AND LEARNING

SHU-CHIU LIU

Nova Science Publishers, Inc.
New York

Copyright © 2009 by Nova Science Publishers, Inc.

All rights reserved. No part of this book may be reproduced, stored in a retrieval system or transmitted in any form or by any means: electronic, electrostatic, magnetic, tape, mechanical photocopying, recording or otherwise without the written permission of the Publisher.

For permission to use material from this book please contact us:
Telephone 631-231-7269; Fax 631-231-8175
Web Site: http://www.novapublishers.com

NOTICE TO THE READER

The Publisher has taken reasonable care in the preparation of this book, but makes no expressed or implied warranty of any kind and assumes no responsibility for any errors or omissions. No liability is assumed for incidental or consequential damages in connection with or arising out of information contained in this book. The Publisher shall not be liable for any special, consequential, or exemplary damages resulting, in whole or in part, from the readers' use of, or reliance upon, this material.

Independent verification should be sought for any data, advice or recommendations contained in this book. In addition, no responsibility is assumed by the publisher for any injury and/or damage to persons or property arising from any methods, products, instructions, ideas or otherwise contained in this publication.

This publication is designed to provide accurate and authoritative information with regard to the subject matter covered herein. It is sold with the clear understanding that the Publisher is not engaged in rendering legal or any other professional services. If legal or any other expert assistance is required, the services of a competent person should be sought. FROM A DECLARATION OF PARTICIPANTS JOINTLY ADOPTED BY A COMMITTEE OF THE AMERICAN BAR ASSOCIATION AND A COMMITTEE OF PUBLISHERS.

LIBRARY OF CONGRESS CATALOGING-IN-PUBLICATION DATA

ISBN: 978-1-60692-786-1

Available upon request

Published by Nova Science Publishers, Inc. ✛ New York

Contents

Preface		vii
Chapter 1	Introduction	1
Chapter 2	Literature Review	3
Chapter 3	Students' Understanding of the Basic Thermal Concepts: Methodology	9
Chapter 4	Students' Understanding of the Basic Thermal Concepts: Results	15
Chapter 5	The Science of Heat in Textbooks	45
Chapter 6	Issues Regarding Teaching Methods and Sequences	65
Chapter 7	Conclusion	69
References		75
Appendix I:	List of the Textbooks in the Study	79
Appendix II:	The 2nd-Phase Questionnaire for German Students	81
Index		87

PREFACE

It is generally accepted that teaching and learning material and activities should be grounded in a deep understanding of students' ideas and the subject matter to be taught. In the domain of thermodynamics, a great deal of research has been done to determine students' (mostly at the primary and middle school levels) misconceptions about heat and temperature; a smaller number has been done to analyze the subject matter as presented in the science classroom. What is very scarce is research which brings together the two aspects to induce a well-grounded instructional reexamination and reconstruction.

The present study is thus conducted to examine secondary students' understanding of basic thermal concepts, such as heat and temperature, along with a conceptual analysis of the scientific knowledge as presented in textbooks. The purpose of the study is to provide a knowledge base for instructional design including features and relations of what students think and what is presented to them, so that more efforts can be made to remove errors and bridge gaps. The study hence consists of two elements: (1) the examination of secondary students' (ages 15-18) understanding of basic thermal concepts, (2) the analysis of textbook accounts. These two elements are finally brought together for discussion and some instructional suggestions are provided. While the data collection is carried out in two countries, Taiwan and Germany, the elicited information has an additional reference to the cultural background, and thus the implications of the study are strengthened. As the results indicate, although the students demonstrate a familiarity of a number of thermal concepts, such as heat, internal energy, temperature, heat transfer, radiant heat, etc., erroneous, confusing and disconnected knowledge is frequently found in their accounts. Most evident is that heat is often explained associated with hotness, temperature, internal energy and radiant heat without a sufficient distinction. Their confusion of these basic concepts leads to wrong interpretations of thermal phenomena and misuses of thermal principles. Textbook analysis also reveals problems regarding accuracy, consistency and coherence in their presentation of the scientific knowledge under study. For example, while heat can be explained correctly in several ways, they pay little attention to clarify from what perspective and in what context an explanation is given. Students' poor understanding of the basic thermal concepts thus seems to find its root or support in the teaching and learning material. The instructional meanings of the results are discussed with a view towards promoting a real understanding of the subject matter through reconstructing the science of heat.

Chapter 1

INTRODUCTION

Two fundamental learning principles, among others, have been recognised by educational researchers: first, new understandings are constructed on the basis of existing understandings and experiences; Students bring to the classroom their own ideas about how the world works and shape their new understandings using these ideas. Their learning of new ideas would not succeed if their initial understanding is not engaged. What may happen is that they simply memorise what is taught to them in order to pass the test while they continue to make sense of the world based on their initial understandings and experiences. The second principle is concerned with real understanding which is expected to occur in the learning process. A real understanding goes beyond factual knowledge, and must contain three elements:

1. A deep foundation of factual knowledge,
2. A conceptual framework which accommodates the learnt facts and ideas,
3. The competence to revise and apply the organised knowledge to various situations.

It is argued that the decisive difference of experts and novice in their knowledge is the effectiveness of taking in new knowledge - what novices see as separate pieces of information is often seen by experts as organised sets of ideas, and thus experts turn out to have more detailed knowledge in an organised pattern (Donovan & Bransford, 2005). Organised knowledge has more strength in retrieval and application and is more effective in acquisition of new ideas than knowledge in pieces. Instruction must therefore emphasize the organising concepts (core concepts) as well as the facts and ideas for students to construct appropriate conceptual maps. As Bruner (1966) has long argued, "A theory of instruction must specify the ways in which a body of knowledge should be structured so that it can be most readily grasped by the learner" (p.41).

Bringing together these two principles, we may argue that science instruction for real understanding must be grounded in the profound knowledge of (1) what and how students think in order to engage students' initial understanding, and (2) the subject matter to be taught in order to clarify facts and the organising concepts. The subject matter must be carefully examined and organised in such a way that students may reflect on their initial understanding and grasp the underlying meanings of scientific knowledge presented to them.

Thermodynamics is no doubt one of the most difficult subjects in science, and an important component of school physics syllabuses in many countries, including Taiwan and Germany. It is expected that students build on their knowledge about basic thermal concepts such as heat and temperature, and demonstrate a sufficient understanding of the mechanism of heat transfer and the ability to explain relevant thermal events from a microscopic as well as macroscopic perspective. Numerous research reports have acknowledged students' difficulties in learning this subject and their misconceptions arising in the learning process, yet relatively little attention has been paid to secondary students, nor to the connections between their conceptual understanding and the actual meaning and presentation of the subject matter.

The present study is thus designed to discover such connections in order to provide a sound knowledge foundation for instructional design and teaching of basic thermodynamics, or the science of heat (to emphasize heat as the core concept). It hence consists of two elements: (1) the examination of secondary students' (ages 15-18) understanding of basic thermal concepts, (2) the analysis of textbook accounts. These two elements are finally brought together in discussion and some instructional suggestions are provided. The data collection is carried out in two countries, Taiwan and Germany, and, therefore, the elicited information has an additional reference to the cultural background and thereby the meaning of the study is strengthened.

The purpose of the study is:

1. To examine secondary students' understanding of basic thermal concepts, such as heat, temperature, heat transfer and thermal equilibrium;
2. To clarify the scientific meanings of these concepts as presented in textbooks and their connections to students' conceptual understanding;
3. To provide science curricular developers and teachers a knowledge base, which should contribute to promoting "teaching for real understanding."

Chapter 2

LITERATURE REVIEW

STUDENTS' CONCEPTIONS IN BASIC THERMODYNAMICS

Much research has been carried out to discover students' alternative conceptions in the domain of thermodynamics and the ways in which these conceptions evolve and change. Among these studies, the concept of heat and temperature is the central topic. It is thus well acknowledged that students hold various intuitive ideas about heat and temperature (a brief summary as shown in Table 1), and that everyday experiences play an essential role in forming these ideas. Most evident is students' confusion between heat and temperature, as research shows that, even after some years of instruction, students do not distinguish well between heat and temperature when they explain thermal phenomena. Their belief that temperature is the measure of heat is particularly resistant to change.

THE NATURE OF HEAT

Everyday, students are exposed to the colloquial term "heat" as a noun, verb, adverb, and adjective, and these multiple uses may lead to confusion (Erickson & Tiberghien, 1985; Romer, 2001; Tiberghien, 1980). Previous studies have provided evidence that young students typically hold substance-based conceptions, such as heat as a substance, something like air or steam (Albert, 1978; G.L. Erickson, 1979), or a substantive fluid, which, for example, "leaked" out to another end while the object is heated on one end (Erickson, 1979;

Erickson, 1980). Ideas as such can be analyzed and predicted from an ontological perspective (Chi, 1992, 2000).

Table 1. A Short Summary of Students' Ideas

	Students' ideas
Heat	Heat as a substance, like air or steam. (Albert, 1978; Erickson, 1979) Heat as a substantial fluid. (Erickson, 1979; Erickson, 1980) Heat makes things rise. (Erickson, 1979) Heat and cold/coldness as two opposite substances. (Erickson, 1979, 1980) Heat as an intensive quality of a body. (Wiser, 1988)
Temperature	A measure of the mixture of heat and cold inside an object. (Erickson, 1979) Temperature as a property of material. (Erickson & Tiberghien, 1985) Temperature is subject to the size of the object and the amount of stuff present. (Erickson, 1979) Temperature measures or quantifies heat. (Kesidou & Duit, 1993) Temperature has a similar meaning to heat. (Erickson, 1979; Kesidou & Duit, 1993; Wiser & Amin, 2001)
Heat and material	Color, thickness and hardness are associated with conductivities. (Clough & Driver, 1985) The speed of the heat movement explains different conductivities. (Clough & Driver, 1985; Tiberghien, 1980)

It is also frequently argued by children that heat makes things rise (in relation to how thermometer works) (Erickson, 1979). Many children mentioned the existence of cold as an opposing substance to heat, and often believed that all objects contain a mixture of heat and cold (Erickson, 1979, 1980). An investigation with 9th-graders reveals that heat as an intensive quality of a body which can be measured by thermometer is a popular idea among the participants (Wiser, 1986, 1988). They believe that heat is "possessed" by objects or systems and do not really understand the nature of heat as an extensive quality which transits between objects or systems.

HEAT AND TEMPERATURE

Temperature is another difficult concept in the thermal science. It has been repetitively reported that young students have difficulties in measuring and quantifying temperatures (Appleton, 1985), and often believe temperature to be a natural property of an object; therefore, some things are naturally hotter while others colder (Erickson & Tiberghien, 1985; Tiberghien, 1980). Even after instruction, students do not always give up their naive idea that some substances, e.g., flour, sugar, or air, cannot heat up (Tiberghien, 1985) or that metals get heated quickly because they attract or hold heat well (Erickson & Tiberghien, 1985). Many students believe that different materials in the same environment have different temperature if they feel different (for example, metal feels colder than wood). This is sometimes connected to the belief of "cold" as a substance opposite to heat; as an example, a middle-school student in Clough and Driver's study (1985) claimed that "The metal is colder because cold passes through it much quicker than the plastic". Many students also believe that the size, the amount of the stuff inside, the surface area, etc. would influence the temperature of the object; for example, the larger ice tube is colder than the smaller one and in turn melts more slowly (Appleton, 1985; Driver & Russell, 1982; Erickson, 1979; Stavy & Berkovitz, 1980).

Abundant research evidence is provided that students have great difficulties to differentiate between heat and temperature. A robust misconception to be noted is temperature as a measure of heat. It is held true by many students that temperature measures or quantifies heat, and, more precisely, an object with higher temperature has more heat (Kesidou & Duit, 2003). Several studies have shown that even after some years of instruction, students still have difficulties to differentiate between heat and temperature when they explain thermal phenomena (Kesidou & Duit, 1993; Tiberghien, 1983; Wiser, 1988). A number of students, for example, in Wiser's study (1988) contended that one object gives out some temperature to another in the process of heat transfer. Although this explanation erroneously replaces heat by temperature, it shows some degree of coherence with the everyday observation. Wiser concludes that students' confusion between heat and temperature leads to further misconceptions of thermal equilibrium. Studies also show that the concepts of heat and temperature are not only challenging to students but also to scientists, who, while making more accurate predictions than students, have difficulties explaining everyday phenomena in interviews (Lewis, 1996; Lewis & Linn, 1994), and maintain divergent representations in their writings (Tarsitani & Vicentini, 1996).

HEAT TRANSFER

As students often consider "heat" and "cold" as two opposing substances acting in thermal events, they do not explain the process of heating and cooling in terms of heat being transferred (Tiberghien, 1983; Tomasini & Balandi, 1987). Some think that the colder object gives "cold" to the warmer object, while others believe that "heat" and "cold" are being exchanged in the same time. Students do not always explain heat-exchange phenomena as interactions. They believe, for example, objects cool down or release heat naturally, without being in contact with a cooler object (Wiser, 1986). As Clough and Driver reveal in their study (1985) that students tend to draw on various "natural observable properties of the materials, such as colour, thickness and hardness" to "explain" different conductivities; for example, a 12-year-old commented that "Plastic grips are softer (than the metal) so they feel warmer". The most popular explanatory framework they discover among the students is that different speed of the movement of heat explains different conductivities; more precisely, a better heat conductor can "attract" or "pull" heat more easily than the others, or else, the less good conductor will "let out" heat more easily. Although secondary students have learnt the molecular basis of heat transfer, they seem to have a poor understanding of it (Kesidou & Duit, 1993; Wiser, 1986).

CONCEPTUAL DEVELOPMENT AND TEACHING METHODS

Studies have also examined the actual processes of conceptual development within the domain. Some early work claims that students' conceptual development parallels historical development of the same concepts (Wiser, 1988; Wiser & Carey, 1983), following which more recent work is presented (Cotignola *et al.*, 2002). Base on their empirical results Wiser and Carey further contend that early scientists and students share similar theories about thermal phenomena that are coherent but incommensurable with the currently accepted scientific theories that differentiate heat and temperature. Wiser's recent work focuses on features of the ontological differences between students' and scientists view and ways in which the shift between ontological commitments is induced (Wiser & Amin, 2001). It is also claimed in this work that students' conceptual change is both evolutionary, as accumulation of information (Gunstone & Mitchell, 1997), and revolutionary, as kind of theory change (Carey, 1986). In regard with the teaching strategy, it is suggested to "encourage students explicitly to differentiate between the scientific and everyday views and then integrate them into a coherent account including both views" (p. 353) and to develop teaching material and methods for "changing ontological commitments" (p.354).

In terms of evolutionary versus revolutionary change, Harrison *et al.* (1999) investigate students differentiating heat and temperature and come to the conclusion that conceptual change in their study was (1) "cumulative and piecemeal" rather than revolutionary as the students "struggled to accommodate new, and for them, counterintuitive ideas" and (2) a gradual conceptual exchange process, as discussed by Hewson and Hewson (1992). These findings correspond with some others in thermodynamics (e.g., Jones et al. (2000) and Lubura et al. (2002)). In examining the leaning process, some research is conduced to investigate the possible role of culture and family in learning about thermodynamics (Jones *et al.*, 2000; Lubben *et al.*, 1999; Slone *et al.*, 1996).

INFLUENCING FACTORS TO STUDENTS' MISCONCEPTIONS

It is argued that everyday experience plays an essential role in students' constructing alternative conceptions. As Clough and Driver (1985) contend, students are familiar with touching and feeling cold/hot in their everyday life, such as the experience of metal feeling colder than plastic, and thus understandably consider temperature as a physical property of material. Also, due to the distinctive different feel of hotness and cold, students may come to the plausible, though incorrect, conclusion that cold is an existing substance opposite to heat (Wiser, 1986). Similarly, Erickson and Tiberghien (1985) point out that the everyday experience of heat resources such as stove, bulb, hot water, etc. brings about the intuitive idea that heat is something that moves within a body or from one body to another.

The everyday use of the term "heat" seems to as well contribute to students' misconceptions. It can be used as noun (e.g, "to keep the heat inside") and as a verb (e.g, "to heat up"), and associated with temperature in some cases and heat quantity in others. Especially when heat is used as a noun, it often implies a substance-like quality, e.g., "keep the heat of the stove on," "heat travels through the metal stick," and "it gains (or loses) heat." It is thus little wonder that students tend to confuse its meanings (Wiser, 1986, Erickson & Tiberghien, 1985).

Summers (1983) concludes from his analysis with several textbooks that alternative conceptions of heat are present in the teaching-learning material. Heat and internal energy are, for example, often used simultaneously without distinction, and heat is sometimes described as if it is a substance. He specifically suggests to replace "heat" with "heating" for heating notifies a process. This is however criticised by Mak and Young (1987), who argue that "heating" can mean the increase in temperature, and thus cause misconceptions. What they suggest instead is to use "heat flow," which, they argue, indicates the changing state of heat. The textbook analysis carried out by Leite (1999) also reveals conceptual or language contradictions and instances of incorrectness. It is especially pointed out by Gotignola et al. (2002) that many popular textbooks fail to make a clear distinction between internal energy and heat. While the terminology and definition of heat is not agreed upon by the science educators, the teaching of this concept can consequently bring about confusions and misconceptions. Furthermore, as reported by Veiga at al. (1989), language used by teachers in their classes sometimes implies incorrect ideas, which reinforce students misconceptions, such as heat as a substance and heat as a fluid. As studies show, students' intuitive and erroneous ideas seem to find their root and support in everyday and learning situations. Therefore, a substantial re-examination of teaching material and methods in regard with these influencing factors should be of significance.

It should be noted that studies on students' concepts about heat have been generally focused on primary and middle school students. It is argued that secondary students do not exhibit their alternative conceptions in an obvious manner for they "are relatively quick at learning verbal labels and scientific-sounding phrases, yet the classroom interaction is normally not long enough to reveal what kind of understanding lies behind such words or phrases" (Clough and Driver, 1985, p.181). Also, there seems to be a lack of varieties of testing instrument in diagnosing students' understanding of a specified concept or theory, while interviewing is commonly used among these studies. The study is thus intended to examine, by means of an alternative testing instrument, secondary students' understanding of the basic thermal concepts such as heat, temperature and thermal equilibrium, which they already learnt through formal instruction. Moreover, the present study seeks to extend the scope by bringing together students' ideas and scientific ideas as presented in the teaching material and thereby reconstructing subject matter for better teaching and learning.

Chapter 3

STUDENTS' UNDERSTANDING OF THE BASIC THERMAL CONCEPTS: METHODOLOGY

METHODS

The methodological design of the investigation is inspired by mixed methods research, which takes "pragmatism" as its worldview or paradigm (Tashakkori & Teddlie, 2003). Within a pragmatic approach, more concern is taken for the research question than for a particular philosophical framework. The pragmatic approach argues that researchers should feel free to ask questions which interest them and to develop methods for use to answer the question. This paradigm allows research to draw on many ideas, including employing "what works", using diverse approaches, and valuing both objective and subjective knowledge (Creswell & Clark, 2007). Among several used in mixed methods research, this study incorporates two approaches:

The use of a qualitative strategy along with a quantitative strategy – This approach involves the adding of open-ended questions to a quantitative research instrument in order to move past the quantitative score generated by the questionnaire and to investigate respondent's own words regarding the particular concept under investigation.

The use of the findings from one study to develop instruments used in another – In this approach, a qualitative study is done to investigate what students think and how they express their thoughts. Afterwards, the extracted statements are used in the development of a standardized questionnaire.

The methods employed in the study include thus a main written test in which students can choose statements as well as write their own, while a previous test with open-ended questions is conducted for developing statements for choice in the main test. In this way, qualitative (textual) and quantitative (choice of "true" or "false") results are embraced, and, moreover, the credibility of the results from the main test is enhanced for its items are based on students' own ideas and wordings.

The first test uses an open-ended questionnaire to elicit students' general ideas about heat and temperature and their predictions and explanations of several familiar thermal phenomena, e.g., the conditions for two objects in contact to transfer heat. As already stated, the purpose of the first test is to derive response items based on students' view and wording for the main study as well as to revise the questions. Therefore, the revised questionnaire should contain useful and legible questions and response items based on students' own thoughts and words. This evaluation phase results in a revised questionnaire with eight multiple-choice questions and an open-ended question; in the former, five response items are given, whereas the latter is intended to strengthen the information with more personal accounts.

Instead of using the common response format: a series of "yes" and "no" items, the questionnaire provides response items which are neither exhaustive nor exclusive as students can select multiple responses or enter their own answer under "other: _____". This means some narrative text will be produced for analysis. It is a fast and easy way to obtain the information of the concepts "accepted" by the student. In the beginning of the test, students are told to tick whichever items stand for their ideas. They are also encouraged to explicate their ideas through drawing. The open-ended question allows respondents to provide their own answers. This gives them opportunity to express their own thoughts and is intended to strengthen the results from the multiple-choice questions. The combination of multiple-choice (statement-based) and open-ended questions as such is intended to provide more complete picture by noting a general trend as well as in-depth knowledge of participants' perspectives (Creswell & Clark, 2007).

PARTICIPANTS

Participants of the investigation are secondary students in Taiwan (n=252; Grade 10-12; age 15-18) and in Germany (n=130; age 15-19); the former are selected from two senior high schools in two satellite cities around Taipei, while the latter from two *Gymnasien* (comparable to English Grammar Schools) in a northern city. These schools represent the common school environment and student population in their countries. Given the fact that students above Grade 11 or 12 should choose their focus subjects, this study deliberately includes students who take science and not because the topic under study is supposed to be learnt by all prior to the time when science becomes optional. It should be noted that a different group of students in each country took part in the previously conducted test for developing the questionnaire with response items. These students are selected from different schools and thus shall not have interplay with those involved in the main investigation.

QUESTIONS

The questions are developed to detect students' understanding of the basic thermal concepts which center on heat. Some questions are factual (the "what" questions), while others are conceptual (the "how" and "why" questions); Some questions concern general meanings, while others involve contextual problems. The questions, for example, "What is heat?" and "What is the function of thermometer?" are context-independent and intended to elicit the respondent's general ideas, whereas questions as the following are contextual-based and intended to discover how students apply their learnt knowledge to a familiar problem.

"Suppose A glass has 200g of water, 25 degrees, and B glass has 50g of water, 90 degrees. What can be concluded based on this information?" (illustration provide in the questionnaire)

"What happens to particles of the metal stick being heated on one end?" (illustration also provided)

Totally nine questions - eight multiple-choice questions and one open-ended question - are included in the questionnaire. These questions address several thermal topics including heat, temperature, heat transfer, phase transition, particle model, etc. They involve, wherever a context is given, only familiar objects and situations, and few mathematical applications in order to avoid confusion and to really look into students' conceptual understanding. The response items are developed separately in the two countries, while the questions are kept identical, which means, all participant students respond to the same questions in the same order, but the response items are different between the questionnaire for the German and Taiwanese students. An exemplar question and its response items for each student group are as follows:

Question 8. What happens to particles of the metal stick being heated on one end?

[Response items for Taiwanese students:]
Temperature rises; more and more particles have heat energy.
Temperature rises; particles move faster and away from each other.
Temperature rises; particles move away carrying heat energy, which follows in from the source.
Temperature rises; particles give heat energy to neighbors by radiation.
Other. _____
[Response items for German students:]
Particles start to move, stretch out and move towards the other end.
Particles take in heat and give a part to the neighbors.
Particles start to move, rub against each other and thereby generate heat.
Heat comes in more and more and gradually outweighs cold.
Other. _____

In the open-ended question students are asked to explain the reasons for ice melting and the differences between the particles in the melted and still solid parts. It is intended to further explore students' understanding of basic thermal concepts including heat, temperature and heat transfer in the case of phase transition and from the microscopic perspective by eliciting information in their own words. It also serves to re-examine students' responses to the multiple-choice questions and to possibly fill in gaps which may arise from the "true" or "false" type of answers.

DATA ANALYSIS

The prior test with only open-ended questions to the main investigation is administered in order to generate another questionnaire including items close to students' view. Thus, its data analysis is conducted for the purpose of finding common responses and wordings. Students' responses are summarized, categorized and selected according to their frequency and clearness in presenting a specific view. Thus, the items included in the main test present ideas that are familiar, or at least easy to comprehend, for the student.

The collected data from the main test include

1. The selected items,
2. Written accounts with the choice of "other" and
3. Written accounts for the open-ended question.

The first two groups of data are elicited by the same questions – multiple-choice questions - and thus are analysed as a set, while the other is treated separately using qualitative analysis technique. The accuracy and implications of each statement as chosen or written by the student are further examined based on a grounded theory approach, whose basic idea is to read (and re-read) the textual data and "discover" or label variables (typically called categories) and their interrelationships. The categories of the written accounts attached to the choice of "other" are treated as variables besides the selected items for quantitative analysis. All these variables are further analyzed using the SPSS package. Consequently, not only qualitative but also quantitative results are produced from the data collection and analysis.

These results are presented according to the conceptual topics the questions are associated with, including

Definition of heat,

1. Process and mechanism of heat transfer, including the conditions for heat transfer between two objects in contact and reasons for metal feeling colder than wood,
2. Particle model in the context of phase transition, including reasons for phase transition and differences between particles in solid and in liquid phase.

It should be noted that choosing item "other" without providing any statement is identified as "no response" in the analysis. Responses which are either illegible or do not match the question they are intended to answer are grouped as "other."

Chapter 4

STUDENTS' UNDERSTANDING OF THE BASIC THERMAL CONCEPTS: RESULTS

The results reveal a diverse meanings of the heat concept, among which are those in relation to (rising) temperature, hotness and internal energy (or thermal energy) most evident. It seems that the familiar image of hot entities, such as fire and the sun, imposes a strong linkage among heat, heat source and heating. Students' accounts as elicited in the investigation also indicate their familiarity with a number of basic thermal terms: heat (energy), temperature, heat transfer, thermal equilibrium, specific heat, conductivity, latent heat, etc. However, these accounts show various levels of disintegration and confusion. Many terms are used in an incorrect and confusing manner. It seems that many students fail to grasp the underlying meanings of the concepts and thus simply memorize what is taught to them.

Although heat and temperature are core concepts in basic thermal science, many students involved in the study do not manage to distinguish one from another. It is often believed that heat is a function of temperature: A body with higher temperature has more heat. As heat is often paralleled to hotness, temperature is thus conceived by the students as the degree of "heat and cold." The ideas about heat, hotness and temperature seem to be intertwined.

The results also indicate that the idea of heat as substance does not cease to act in secondary students' conceptualization of the thermal phenomena. Such idea is not presented directly in the factual question such as "What is heat?" but becomes apparent when the student explains how heat is being transferred. It seems that the student is familiar with the kinetic view, but turns back to the material view when a more detailed account is required. Moreover, as believed by a number of the participants, heat acts in a similar way like electricity and can be transferred only if the objects are "conductors." As the data show, the majority of the participants only understand the mechanism of heat transfer at a superficial level. They do not realize that objects, regardless their material and size, would come to the state of thermal equilibrium with their environment.

It is evident that the students often articulate the microscopic states and actions based on the macroscopic phenomena. The learnt concept of friction heat, for example, is being used to predict what happens when particles move in order to transfer heat, as students do not distinguish the context of these events. The reconciliation of particle motion and friction heat is manifestation of the confusion between the microscopic and macroscopic systems.

In the following these results will be elaborated in three sub-sections:

1. Definition of heat
2. Process and mechanism of heat transfer
3. Particle model in the context of phase transition. The tables illustrated in this section contain students' response items in a shortened form. The marked area in the tables contains items which are provided in the questionnaire. Other items listed in the tables are categorized from responses in "other" with a rate of 5% and beyond.

DEFINITION OF HEAT

Students' responses show a diversity of meanings of the heat concept. These meanings are to various extents associated with other thermal terms and concepts, e.g., temperature, hotness/cold, particle motion and internal energy, and also some mathematical applications (Figure 1). The distinctions and relations among these representations are, however, often ambiguous and erroneous. Most obvious is the evidence that heat is often related to higher temperature, something hot, and a kind of energy inside a body. Heat as "a kind of energy making temperature rise," "the rise of temperature and the expansion of the body" and "hotness; the opposite to cold," are popular notions among the students under study. In many cases, the definition of heat is a mixture of heat source, hotness and internal energy. A number of students, for example, voted for the notion that heat is "a kind of energy released exclusively from hot things such as fire." It also reoccurs that students explain heat in terms of particle motion and mathematical representations, yet the accuracy of such accounts appears to be poor. It is worth noting that some students, exclusively from Germany, confuse heat with entropy, believing that "Heat is entropy, which is a function of mass and temperature".

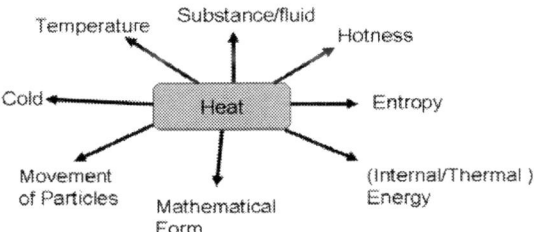

Figure 1. Students relate heat to various things.

The idea of heat as substance, in particular material fluid, is not directly addressed by the students under study. It comes, however, often to the surface when the question involves the application of the learnt knowledge. The student may state that heat is a kind of energy but still fall back to the concept of heat substance when a more detailed account of a thermal event is required. This indicates that the concept of the material-like heat still finds its position, and, moreover, forms part of a "hard core," in the student's conceptual framework (as previously argued by Niaz (2006), based on his study with science major freshman students).

HEAT AND HOTNESS

The data show that nearly half of the Taiwanese students relate heat to rising temperature ("a kind of energy making temperature rise") and one out of three confuses heat and temperature ("degree of hotness and cold"), as illustrated in Table 2. Similarly, a majority of the German students define heat as "the rise of temperature and the expansion of the body", and meanwhile one out of four holds true that heat is "hotness, an opposite to cold" (see Table 3). It seems that the image of familiar heat sources such as fire and the sun imposes a strong linkage among heat, heat source and heating. This linkage is also illustrated in another frequently chosen item which states that heat is "a kind of released energy from something hot" (see Table 2).

To refer to "cold" and "heat" as two contrasting states or qualities in describing the phenomena of ice melting seems common in the students' responses, especially among the German sample. The students often associate fixed particles with cold and moving particles with hotness or attribute the hardness of ice to cold and the looseness of water to heat. The following extracts of students' responses can illustrate this idea:

> "..the small particles, which were due to cold held together, become loose due to heat." (German 10th-grader)

> "Cold makes [things] fixed and heat makes [them] moving." (German 10th-grader)

> "Those [particles] in the cold concentrate themselves, whereas those in the warmth/heat stretch out." (German 10th-grader)

Table 2. Taiwanese Students' Responses to "What is Heat?" 252 cases; 3 "no response"

		Responses		Pct of Cases
		Count	Pct	
What is heat (a)	A kind of material; a body is hotter if having more of it	9	3.0%	3.6%
	A kind of energy making temperature rise	115	38.3%	46.2%
	A kind of energy released exclusively from hot things such as fire	38	12.7%	15.3%
	A quality of the material; its degree of hotness or cold	81	27.0%	32.5%
	A kind of energy	29	9.7%	11.6%
	Other*	28	9.3%	11.2%
Total		300	100.0%	120.5%

A German student even believed that ice starts to melt because its cold is taken away. In his words, ice is "the state of cold", and "when we take cold away", the ice changes its state. It is also worth noting that some students conceived of ice melting as a process of exchanging "cold" and "heat". One German 11th-grader commented that "due to the warm temperature ice loses the stored cold and take in heat of the room", while another put it this way: "the room withdrew cold from the ice". This concept is elaborated microscopically by one of the German students from Grade 11. According to him, ice contains particles with "cold" and it melts when its particles start to exchange with those with "heat" in the environment. His illustration, as Figure 2, can serve to visualize this concept. Despite its incorrectness, this concept provides a, to some extent, self-sufficient and coherent explanation.

Table 3. German students' responses to "What is Heat?"

		Responses		Pct of Cases
		Count	Pct	
What is heat (a)	Entropy; a function of mass and temperature	11	4.5%	8.5%
	Movement of particles, and its consequently produced energy	113	46.3%	86.9%
	Hotness, opposite of cold	34	13.9%	26.2%
	Rise of temperature, and expansion of the body	80	32.8%	61.5%
	Other	6	2.5%	4.6%
Total		244	100.0%	187.7%

130 cases
a Question 1
including notions such as "a kind of energy change during a change in temperature," "a kind of energy released by everything," "a kind of energy causing temperature rise or change in state" and "a kind of energy which everything has." Each contains no more than 5% of Pct of Cases.

HEAT AND TEMPERATURE

Although heat and temperature are core concepts in basic thermal science, many students after years of instruction do not manage to distinguish one from another. As illustrated in Table 2, nearly one out of three Taiwanese students considered heat as "a quality of the material; its degree of hotness or cold". The meaning of heat is confused with that of temperature.

Similar results are revealed in the question about the function of thermometer. It should be noted that a high proportion of the Taiwanese and German sample, with nearly the same rate for each group, showed no difficulty to point out that the function of thermometer is to measure "the temperature of a body and its change" (see Table 4 and Table 5). However, a half of the German students, as seen in Table 5, believed that thermometer can also measure "temperature of two objects, thereby indicate which has more heat than the other". This conforms to the misconception that heat is a function of temperature – generally, higher temperature means more heat. As a number of students agreed, "...if thermometer reads 20 degrees on 20g of water, we know the water has 400 cal of heat" (see Table 5). It is especially concerning that two out of five German students suggested that "heat and cold" can be also measured by thermometer, as shown in Table 5. Heat and cold are again seen as two opposite qualities that can be quantified. It seems that the meaning of temperature as the degree of hotness and cold is being reinterpreted as the degree of "heat and cold." To note, heat is for many students a similar term to hotness.

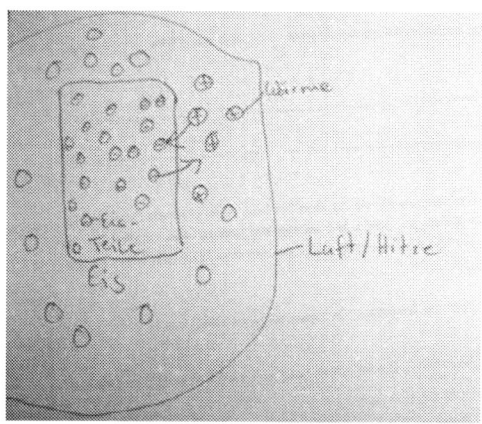

Figure 2. "Cold"(-) and "heat"(+) are exchanging when ice gets in touch with warm air.

Mathematical formula is frequently used by the students under study. However, their meaning is rarely clarified. For example, a Taiwanese student cited that thermometer can be used to measure a body's change in temperature, ΔT, and in turn ΔH can be calculated by $\Delta H = M*S*\Delta T$. No further explanation on either of the terms is given by the student.

Several German students specifically associate heat with temperature in a microscopic level. For example, a German 10th-grader, in responding to the question about ice melting, mentioned that heat is "dependent on the average speed of the particles." While the concept of temperature as "average speed of particles" is presented, the student does not necessarily understand what it really means. For some students, heat transfer is understood as the "exchange" of the particles with different degrees of movement, which means, the hotter system gives its particles with higher speed to the colder system where the particles move at a lower or no speed. For example, a German 10th-grader, in explaining the process of ice melting, stated that "temperature is the movement of the particles." However, he further declared that ice only has static particles, and it melts because the room temperature causes a replacement of the static particles by those in motion. These particles in motion are presumably from the environment. Such account clearly indicates a concept of heat as something like a material fluid, and heat transfer as a real flow of particles. In the case of ice melting, what happens is that the particles with motion (outside at higher temperature) replace those at rest (in ice at lower temperature). While temperature is understood as a function of the speed of the particle motion, this understanding is reconciled with the view of heat substance and in turn forms an alternative conception of heat transfer.

Table 4. Taiwanese students' responses to "What is the function of thermometer?"

		Responses		Pct of Cases
		Count	Pct	
Function of thermometer (a)	Temperature of a body and its change	214	77.0%	86.6%
	Change of the volume, for example, while heating water	8	2.9%	3.2%
	The amount of heat in a body; for example, if thermometer reads 20 degrees on 1g of water, we know the water has 20 cal of heat.	23	8.3%	9.3%
	Temperature of two objects, thereby indicate which has more heat	24	8.6%	9.7%
	Other	9	3.2%	3.6%
Total		278	100.0%	112.6%

252 cases; 5 "no response"
a Question 3

Table 5. German students' responses to "What is the function of thermometer?"

		Responses		Pct of Cases
		Count	Pct	
Function of thermometer (a)	Temperature of a body and its change	111	41.3%	86.0%
	Heat and cold	54	20.1%	41.9%
	The amount of heat in a body; for example, if thermometer reads 20 degrees on 20g of water, we know the water has 400 cal of heat.	28	10.4%	21.7%
	Temperature of two objects, thereby indicate which has more heat than the other	74	27.5%	57.4%
	Other	2	.7%	1.6%
Total		269	100.0%	208.5%

130 cases; 1 "no response"
a Question 3

It is found that the term "temperature exchange" (*Temperaturaustausch*) used by the German students may reinforce the difficulties in differentiating heat and temperature. This term does not have a meaningful equivalent in either English or Chinese, and suggests temperature to be a substance, similar to heat. For example, while explaining the ice melting process, a German 13th-grader commented "A temperature exchange between the ice and room occurs, which means, the two bodies equalize their heat". The exchange of temperature leads to, or is equal to, the exchange of heat. The meaning and usage of this term clearly draws a similarity between heat and temperature, and can understandably cause or enhance the confusion of these two concepts.

HEAT AND INTERNAL (THERMAL) ENERGY

The confusion between heat and thermal energy is demonstrated in the popular statements, such as heat is "movement of particles, and its consequently produced energy" (86.5%; in Table 3) and, implicitly, heat is "a kind of energy released exclusively from hot things such as fire" (15.3%; in Table 2). A German 10th-grader, believing that ice starts to melt because it is heated and thereby its particles start to move (in fact, they only move faster), contended that through the particle motion, "energy or heat is generated" and "that's what the melting of ice brings about". It is apparent that the student does not distinguish heat from thermal energy. For him, what the ice receives to make it melt (heat) and what the particle motion generates (thermal energy) are the same thing.

In discussing the illustration in Figure 3, a half of the German students believed that the "internal heat" can be calculated by dividing temperature with volume. For these students, heat is in direct ratio to temperature, but in inverse ratio to the volume which the heat resides. It seems to be intuitive to think that heat is "distributed" in a bigger volume and "concentrated" in a smaller volume, and thus less intensive in the former than in the latter. In this view, heat is associated with a kind of "intensity" of the hotness.

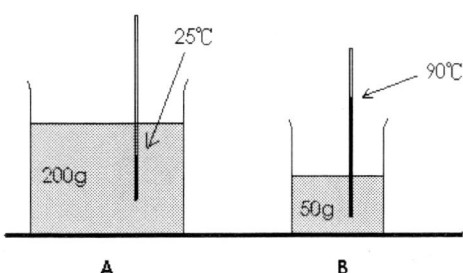

Figure 3. Two cups of water with different temperature and volume.

A number of the students concluded from the same figure that "A has 5000 cal of heat, while B has 4500" (22.7% in Table 6 and 17.2% in Table 7). These students calculated heat by multiplying temperature and mass. It may be a distorted concept from the familiar formula repeated in the basic thermal lesson: $\Delta H = M*S*\Delta T$, where delta is removed and S is valued 1 for water. The students fail to understand the meanings of the formula - first of all, heat is a quality in transition; secondly, heat can be calculated while a change in temperature happens. Thus, this formula is transformed into $H=M*S*T$. It is then little wonder that a conclusion of A with 5000 cal of heat and B with 4500 cal is drawn. It should be noted that nearly one out of five German students voted for the statement that "Mixing A and B, temperature will be balanced while heat will sum up." As heat is often believed to be an internal quality, it is not a surprising conclusion that mixing two cups of water leads to addition of the heat they have. This again indicates that students have a concept of heat which is a function of temperature and is something inside a body. Heat seems to be often confused with internal energy and can be calculated using some simple formula. This also provides a more concrete version of the finding discussed in the previous subsection that heat is often considered to be an internal quality and is a function of temperature.

Table 6. Taiwanese students' responses to
"Suppose A glass has 200g of water, 25 degrees, and B glass has 50g of water, 90 degrees. What can be concluded based on this information?"

		Responses		Pct of Cases
		Count	Pct	
Heat and thermal equilibrium (a)	A has 5000 cal of heat, while B 4500.	56	17.8%	22.7%
	B has more heat than A for it has higher temperature.	10	3.2%	4.0%
	Mixing A and B, their final amount of heat will be between 4500 and 5000 cal.	41	13.0%	16.6%
	Mixing A and B, a final temperature between 25 and 90 degrees will be reached.	200	63.5%	81.0%
	other	8	2.5%	3.2%
Total		315	100.0%	127.5%

252 cases; 5 "no response"
a Question 4

The data show that a similarly high proportion of the German and Taiwanese sample chose correctly the statement that "Mixing A and B, a final temperature between 25 and 90 degrees will be reached" (as shown in Table 6 and Table 7). It does not seem to be a problem for the students to predict what happens to two cups of water with different temperature after they mixed. This question alone may lead to a conclusion that students understand the meaning of thermal equilibrium. However, when the objects involved are of different material, state, and shape (as discussed in the later subsection) it becomes clear that students' understanding of thermal equilibrium is only at a superficial level. They could not apply the concept of thermal equilibrium to diverse phenomena, and thus frequently trust in their intuition to make prediction.

Table 7. German students' responses to "Suppose A glass has 200g of water, 25 degrees, and B glass has 50g of water, 90 degrees. What can be concluded based on this information?"

		Responses		Pct of
		Count	Pct	Cases
Heat and thermal equilibrium (a)	A has 5000 cal of heat, while B 4500.	22	10.3%	17.2%
	With reference to their volume, B has more heat than A (A:25/200<B: 90/50=360/200)	54	25.2%	42.2%
	Mixing A and B, a final temperature between 25 and 90 degrees will be reached.	114	53.3%	89.1%
	Mixing A and B, temperature will be balanced while heat will sum up.	24	11.2%	18.8%
Total		214	100.0%	167.2%

130 cases; 2 "no response"
a Question 4

Also, as *radiant heat* is a part of the science lesson and seems to be compatible with the idea of heat as internal energy, students also tend to give accounts that connect heat to radiant heat as well as internal energy. Comments such as that "Heat energy is released by all substances at over absolute zero." and "(Heat is) the energy that exists inside a body, even for a body at 0°C." (an illustration in Figure 4 is provided along with this statement by a 12th-grader) are found among students' responses to "What is heat?"

While responding to the question of heat transfer between two bodies in contact, several students specifically pointed out that there is no condition for heat transfer as such for "heat energy is being always transferred" or "No matter cold or hot, the both body objects will release heat until they have the same temperature"(German 11th-graders). It is clear at this point that the confusion between heat and radiant heat leads to a further misconception of heat transfer.

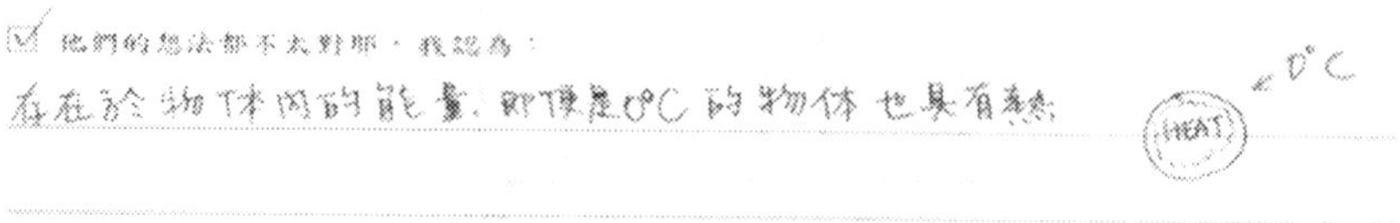

Figure 4. A 12th-grade student pointed out that heat is a quality possessed by the object, even at 0°C.

PROCESS AND MECHANISM OF HEAT TRANSFER

Several students provided accurate microscopic explanation of heat transfer in the case of ice melting. A German student from Grade 10, for example, noted that "The energy content of particles in melted ice [water] is higher than that of particles in fixed ice. Ice keeps melting, until it has the same energy content with the air." Another detailed explanation is given by a student from Grade 11: "The internal energy of the air is transferred to the water [ice] molecules. The water [ice] temperature is approaching the air temperature until the both are the same. The outer parts of the ice, which have contact with air, gets warm first and flow down as soon as the temperature go over the melting point." However, such explanations are exceptions among the responses.

Virtually no students referred heat directly to substance. The statement that heat is a kind of energy is easy for them to recall as repetitively taught. It seems that, however, in describing the movement of heat they tend to visualize it in a corporeal manner, i.e., heat transfer is often depicted as if it is an actual continuous transmission. Many students also maintain a material view of heat. A question which requires students to describe actions of the particles in the metal stick as shown in Figure 5 reveals that more than a third of the Taiwanese participants believed that particles "move away carrying heat energy which flows in from the source" (Table 8). Also, the idea that "more and more particles have heat energy" in this heating process is supported by a number of students and indicates likewise a material view of heat. Similarly, nearly a third of German students agreed with the notion that particles "start to move, stretch out and move towards the other end," and meanwhile two out of five held true that particles "take in heat and give a part to the neighbors" (Table 9). These results indicate that students still remain the idea of particles behaving like a carrier and transporting heat in a manner like a flow (Figure 6). Figure 7 shows a sketch by a 12th-grader where a metal is heated on one end and heat goes around the metal like an "infinite" circuit.

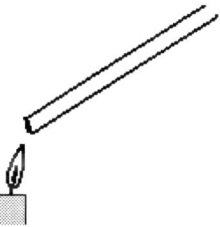

Figure 5. What happens to the particles of the metal stick?

Table 8. Taiwanese students' responses to "What happens to particles of the metal stick being heated on one end?"

		Responses		Pct of Cases
		Count	Pct	
Heat transfer in particles (a)	Temperature rises; more and more particles have heat energy.	35	12.1%	14.1%
	Temperature rises; particles move faster and away from each other.	73	25.3%	29.4%
	Temperature rises; particles move away carrying heat energy, which flows in from the source.	96	33.2%	38.7%
	Temperature rises; particles give heat energy to neighbors by radiation.	46	15.9%	18.5%
	Heat is transported through conduction.	24	8.3%	9.7%
	other	15	5.2%	6.0%
Total		289	100.0%	116.5%

252 cases; 4 "no response"
a Question 6

It is worth noting that a majority of the German students advocated the idea that particles "start to move, rub against each other and thereby generate heat" when the metal stick is being heated (Table 9). Students seem to visualize the behavior of the particles in a macroscopic manner and to equate heat transferred through particle motion with heat generated by friction. The learnt concept of friction heat is being used to predict what happens when particles move in order to transfer heat, as students do not distinguish the context of these events. The reconciliation of particle motion and friction heat is manifestation of the confusion between the microscopic and macroscopic systems.

Table 9. German students' responses to "What happens to particles of the metal stick being heated on one end?"

		Responses		Pct of Cases
		Count	Pct	
Heat transfer in particles (a)	Start to move, stretch out and move towards the other end.	38	18.9%	29.5%
	Take in heat and give a part to the neighbors.	54	26.9%	41.9%
	Start to move, rub against each other and thereby generate heat.	84	41.8%	65.1%
	Heat comes in more and more and gradually outweighs cold.	19	9.5%	14.7%
	other	6	3.0%	4.7%
Total		201	100.0%	155.8%

130 cases; 1 "no response"
a Question 6

Figure 6. A popular idea: The particle carries "heat" (H) and moves like a flow.

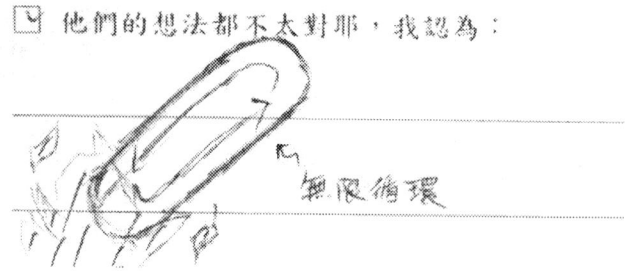

Figure 7. Heat transfer depicted as a current by a 12th grader.

CONDITIONS FOR HEAT TRANSFER BETWEEN TWO OBJECTS IN CONTACT

As Table 10 and Table 11 show, only a half of German and slightly more than a half of Taiwanese students considered temperature difference as a condition for heat transfer between objects in contact. A considerable number of students believed that heat transfer starts to take place when the two objects contain difference amount of heat. It is especially warning that more than a half of German students supported this argument. It indicates that heat is often confused with internal energy, and heat transfer is consequently considered to be a result of difference in heat amount between the systems in contact.

Table 10. Taiwanese students' responses to "Under what conditions does heat move from one object to another when they are in contract?"

		Responses		Pct of Cases
		Count	Pct	
Conditions for heat transfer(a)	Having different temperature; from higher to lower	156	49.2%	62.4%
	Having different amount of heat	65	20.5%	26.0%
	Depends on their ability to absorb or release heat	21	6.6%	8.4%
	Depending on their size, mass, state and so on	66	20.8%	26.4%
	other	9	2.8%	3.6%
Total		317	100.0%	126.8%

252 cases; 2 "no response"
a Question 2

The idea that heat transfer occurs depending on the properties of the objects, such as the size, mass, state, etc. is held true by one out of four Taiwanese students. A German student also supported this idea and explicitly noted: "the first object must be twice as big as the second". Previous studies have given evidence that children tend to believe a linkage between heat transfer and the size and shape of the objects in question, yet surprisingly it is also repeated by secondary students in the present study.

Table 11. German students' responses to "Under what conditions does heat move from one object to another when they are in contact?"

		Responses		Pct of Cases
		Count	Pct	
Conditions for heat transfer(a)	Both objects are capable of transporting heat.	70	25.3%	54.7%
	They have different amount of heat; heat goes from more heat to less.	86	31.0%	67.2%
	They have different temperature; heat goes from higher temperature to lower until they have the same temperature.	69	24.9%	53.9%
	One of them has to have heat (being hot or warm).	49	17.7%	38.3%
	other	3	1.1%	2.3%
Total		277	100.0%	216.4%

130 cases; 2 "no response"
a Question 2

On the side the German sample, the idea that if heat transfer should occur, the objects must be capable of transporting heat is especially popular. For these students, the occurrence of heat transfer is subject to a specified property of the objects - the capability of transporting heat. It seems that heat acts in a similar way like electricity and can be transferred only if the objects are "conductors." This confirms again that students do not fully understand the nature of heat as an energy form and may turn to a description of material-like heat depending on the question.

REASONS FOR METAL FEELING COLDER THAN WOOD

Diverse reasons for metal feeling colder than wood are elicited in the investigation. As Table 12 and Table 13 show, "Wood absorbs (and stores) heat better (than metal)." is the most popular item among both Taiwanese and German students (42.7% for the former and 51.9% for the latter). To follow, "Metal transfers heat faster." and "Metal has smaller specific heat." are frequently suggested by the Taiwanese students as the correct answer (see Table 12). On the contrary, two out of five German students believed that metal feels colder because it "has higher density, so does not take in so much heat", while "Metal is a conductor (and wood is not) and can take in cold" is also a frequently voted item among the German sample, as illustrated in Table 13. It is worth noting that apart from the most accepted item the two sample groups gave very different responses. Transfer speed and specific heat are mentioned frequently by the Taiwanese sample, but rarely by the German. In contrast, density and conductivity are considered to be important factors of the cold feel of metal by the German students, but they received no or relatively little support from the Taiwanese sample. Also, while the German students stated "Metal is a conductor (and wood is not)" their Taiwanese counterparts suggested a modified idea that metal is a better conductor than wood, as some wrote "Metal has higher ability to conduct heat" (see Table 12).

Table 12. Taiwanese students' responses to
"Why does metal generally feel colder than wood?"

		Responses		Pct of Cases
		Count	Pct	
Reason(s) for metal feeling colder(a)	Metal has a lower temperature.	15	5.3%	6.1%
	Metal can absorb heat better.	105	36.8%	42.7%
	Metal has less amount of heat in it.	15	5.3%	6.1%
	Metal is naturally colder.	6	2.1%	2.4%
	Metal transfers heat faster.	46	16.1%	18.7%
	Metal has smaller specific heat.	53	18.6%	21.5%
	Metal has higher ability to conduct heat	16	5.6%	6.5%
	other*	29	10.2%	11.8%
Total		285	100.0%	115.9%

252 cases; 6 "no response"
a Question 5
* including notions such as "Metal has smaller specific heat, so has more change in temperature." "Metal has smaller specific heat, so its temperature lowers easily." "Metal attracts heat from hand faster." and "Metal has bigger specific heat". Each contains no more than 2% of Pct of Cases.

Table 13. German students' responses to "Why does metal generally feels colder than wood"

		Responses		Pct of Cases
		Count	Pct	
Reason(s) for metal feeling colder(a)	Wood absorbs and stores heat better.	67	35.8%	51.9%
	Metal has higher density, so does not take in so much heat.	52	27.8%	40.3%
	Metal is a conductor (wood not) and can take in cold.	40	21.4%	31.0%
	Metal has sensitive and cold particles.	5	2.7%	3.9%
	Metal takes away heat from hand more quickly.	14	7.5%	10.9%
	other	9	4.8%	7.0%
Total		187	100.0%	145.0%

130 cases; 1 "no response"
a Question 5
* including notions such as "Metal transfers heat better, thus cools better." "Metal is a conductor (wood not) and can take in heat." "Metal is more easily affected by heat and cold." and "Metal gives out more heat". Each contains no more than 3% of Pct of Cases.

Although the Taiwanese students seem to give their choice to items "closer" to the scientific explanation, some notes given to the "other" item reveal underlying misconceptions. They pointed out that metal has a smaller specific heat than wood, but did not necessarily understand that heat is therefore transferred faster from metal to the hand than wood to the hand. What they actually understood is that, for example, because of smaller specific heat, metal "obtains [reaches] the room temperature more easily [than wood]" and thus feels colder (10th-grader). A 12th-grader also supported this view by noting that "Metal has small specific heat, so its temperature changes easily," in comparison to wood, and therefore "metal is colder at room temperature (than wood)." It becomes clear that what makes metal feel colder than wood is that it is really colder because "it reaches the room temperature easily". It is evident that the sensorial hotness and cold are not distinguished from the real temperature, and are not understood in terms of the transition of heat. Despite their familiarity with the scientific concept of "specific heat," the students did not draw a correct conclusion.

PREDICTION OF HEAT TRANSFER BETWEEN OBJECTS AND ENVIRONMENT

The following two questions included in the test are intended to reveal students' understanding of the mechanism of heat transfer, especially the concept of thermal equilibrium, through different, but familiar, settings:

- *Question 7: Suppose after placing a metal and a wooden stick under room temperature for a while, we move and leave them outside under the midday sun (approx. 35 degrees). What would happen to them?*
- *Question 8: Suppose after placing a metal plate, a bowl of flour and a cup of water under room temperature for a while, we move and leave them in a freezer. What would happen to them?*

The results in the two sample groups are similar, indicating that the participant students do not apply the learnt concepts to solve the problem, but often gave intuitive responses instead. Evidently, many students did not believe that metal and wood would reach the same temperature at the end if they are placed together under the midday sun. A considerable number of the students voted for the idea that metal will have more temperature rise than wood (see Table 14 and Table 15; 60.6% of the Taiwanese and 52.3% of the German). Only a quarter of the participant students believed that "The both temperature increases and will reach the same degree", as Table 14 and Table 15 show. It is worth noting that this statement contains nearly the same rate in the both sample groups.

It is frequently suggested by the German students that, as shown in Table 15, "Metal is heated faster and so contains more heat than wood after a while." (77.7%) and "Metal is heated faster, but cannot keep heat as long as wood" (62.3%). For these students, metal attracts heat faster, so "contains" more heat, but cannot "store" heat as well as wood. It seems that "losing heat easily" is interpreted as an inability to "keep" heat inside. A number of the Taiwanese students also believed that "metal will possess more heat than wood" at the end. These data indicate again that students' idea of the "internal" heat is predominantly involved in their explanation of thermal phenomena as phrases such as to "contain" and "keep (or store)" heat are frequently used. Notably, to derive the conclusion that metal will eventually contain more heat than wood based on the fact that metal absorbs heat faster is plausible, but truly intuitive.

Table 14. Taiwanese students' responses to "What happens to the metal and wooden stick being moved from room to outside under the midday sun?"

		Responses		Pct of Cases
		Count	Pct	
Thermal equilibrium with environment (a)	Metal's temperature will rise; wood changes little or not at all.	151	57.4%	60.6%
	The both temperature increases and will reach the same degree.	64	24.3%	25.7%
	Wood can store temperature better, so will finally have higher temperature than metal.	9	3.4%	3.6%
	They'll reach the same temperature, but metal will possess more heat than wood.	24	9.1%	9.6%
	other	15	5.7%	6.0%
Total		263	100.0%	105.6%

252 cases; 3 "no response"
a Question 7
* including notions such as "They'll finally have the same amount of heat." "They'll both become warmer." "Their change in temperature is subject to mass." and "They'll receive the same amount of heat". Each contains no more than 2% of Pct of Cases.

As the data show, German students' responses are consistent in these two questions, as similar, but low, rates are revealed for the correct statement that objects will approach the same temperature (25.4% in Table 15 and 26.2% in Table 17). In contrast, their Taiwanese counterparts tend to predict what happens depending on the situation and to overlook the common principle that underlies the two questions. For many, metal and wooden stick will not come to the same temperature under the midday sun, but it is more likely that metal, flour and water will do in the freezer (see Table 14 and Table 16). A considerable number of Taiwanese students held true that the objects of different material in the freezer will all "lose heat, until they contain the same amount of heat" (as illustrated in Table 16). It is an alarming fact that the percentage of the students choosing the correct notion is surprisingly low in the both sample groups. The majority of the students believed that metal will have the most temperature change (e.g. 60.6% in Table 14), or eventually contain most or least heat (e.g. 52.3% in Table 17) in the two cases.

Table 15. German students' responses to "What happens to the metal and wooden stick being moved from room to outside under the midday sun?"

		Responses		Pct of Cases
		Count	Pct	
Thermal equilibrium with environment (a)	The both temperature increases and will reach the same degree.	33	11.6%	25.4%
	Metal takes in heat and its temperature rises. Wood changes little or not at all.	68	23.9%	52.3%
	Metal is heated faster and so contains more heat than wood after a while.	101	35.6%	77.7%
	Metal is heated faster, but cannot keep heat as long as wood.	81	28.5%	62.3%
	other	1	.4%	.8%
Total		284	100.0%	218.5%

130 cases
a Question 7

It seems to us that these students fail to understand the fundamental thermal principles: (1) Heat flows from higher to lower temperature; (2) Heat transfer does not stop until the systems are in the state of thermal equilibrium where they have the same temperature. As a result, they reply to the related questions often based on their intuitive concepts. It seems to be more plausible for the students that metal will become hotter than wood when the both are heated by the same source and likewise the metal should become colder than flour in the freezer. Another problem such responses reveal is that the students do not distinguish "feel hot/cold" and "be hot/cold." It is correct to say that metal feels colder than wood, but metal is not really colder than wood. It is indeed warning that after years of science lessons the secondary students under study still apply intuitive ideas to the familiar problems concerning fundamental thermal principles.

Table 16. Taiwanese students' responses to "What happens to metal, flour and water being moved from room to freezer?"

		Responses		Pct of Cases
		Count	Pct	
Thermal equilibrium with environment (a)	All lose heat, until they contain the same amount of heat.	29	10.9%	11.8%
	All objects approach the same temperature as that of the freezer.	114	42.9%	46.3%
	All become colder; but Metal will be the coldest, and flour least cold.	110	41.4%	44.7%
	All become colder; but ice will be the coldest for it feels so.	7	2.6%	2.8%
	other	6	2.3%	2.4%
Total		266	100.0%	108.1%

252 cases; 6 "no response"
a Question 8

PARTICLE MODEL IN THE CONTEXT OF PHASE TRANSITION

As described in the Methodology section, an open-ended question is included to obtain more personal accounts and in turn to enhance the strength of the data. This question (containing two sub-questions), as follows, is intended to elicit information on students' understanding of particle model in the context of phase transition, using the case of ice melting.

- *Question 9: When we take out ice from the fridge, it starts immediately to melt. Please explain (1) why, and (2) what is the difference between particles in still frozen ice and the melted ice.*

Students' responses to this question reveal an evident correspondence to the previously discussed findings. In general, most of the students have in mind a picture of particle model in which particles have different distances and speeds depending on their phases. However, their descriptions indicate an inadequate understanding of particle model as applied in the context of phase transition. Especially noticeable is that they are intertwined with the misconceptions they hold regarding heat and other thermal concepts and that the students often articulate the microscopic states and actions based on the macroscopic phenomena.

Table 17. German students' responses to "What happens to metal, flour and water being moved from room to freezer?"

		Responses		Pct of Cases
		Count	Pct	
Thermal equilibrium with environment (a)	Temperature of all objects approaches that of the freezer.	34	13.7%	26.2%
	All become colder. But metal will be extremely cold, while flour keeps, or changes slightly, its temperature.	95	38.2%	73.1%
	All lose heat. At the end metal has the least heat, and flour the most.	68	27.3%	52.3%
	All absorb cold. Metal can do it the best, water second best, and then flour.	50	20.1%	38.5%
	other	2	.8%	1.5%
Total		249	100.0%	191.5%

130 cases
a Question 8

REASONS FOR PHASE TRANSITION

In general, ice melting as a result of higher temperature or heat in the environment is the most common idea elicited from the corresponding questions. Students in the higher grade can account for this process in a more precise way. For example, it is frequently mentioned among the 12th- and 13th-graders that "ice takes in heat/ heat energy (from the environment)," whereas the reason given by the 10th-graders is often as simple as "due to the higher temperature (of the environment)." Also, more students in Grade 12 and 13 than in Grade 10 pointed out the critical temperature where phase transition starts to happen.

The data show that the Taiwanese students tend to account for ice melting from a macroscopic view. They typically mentioned "heat absorption," and sometimes further explained this as a result of temperature difference between ice and the outside room and its consequent action to reach the state of thermal equilibrium. The following extracts illustrate such responses:

> "...outside temperature is higher, so ice starts to absorb heat in order to reach the same temperature... While the melting point is lower than the room temperature, it starts to melt". (10th-grader)
> "The room temperature is higher than inside the freezer. In order to reach thermal equilibrium ice starts to absorb heat and melt". (10th-grader)
> "The ice, which is taken out of the freezer, starts to melt because the outside temperature is higher. Approaching thermal equilibrium with the outside temperature, it will absorb heat and melt". (11th-grader)
> "The room temperature is higher. It (ice) starts to absorb heat in order to reach thermal equilibrium". (12th-grader)

It should be noted that although the term "thermal equilibrium" is sometimes directly used in their responses, the students do not, however, apply it to the familiar thermal events, as Question 7 and 8 reveal (as discussed in the previous sub-section). Many students seem to know that the concept of thermal equilibrium should be applied in the case of ice melting, but turned to make an intuitive decision in the case of, for example, metal and wood placed in the midday sun. This result indicates that the concept of thermal equilibrium is only understood at a superficial level, where the students may arrive at an intuitive conclusion when the situation fits.

Among the Taiwanese sample, there are few responses relating to particle motion in giving the reason for ice melting. Their notion on the particle model is principally revealed by the sub-question about the differences between particles in ice and in water. Yet, these notions show that their understanding of the particle model is inadequate and often incorrect. For the majority of the Taiwanese students, the particles in ice have higher "density" than those in water. Also, "distance between each other" and "phase" are considered to be what draws difference between them. It seems to be a clear tendency among the students to switch between microscopic and macroscopic descriptions while their linkage is erroneously proposed.

On the side of the German sample, statements in relation to particle model are frequently provided to account for ice melting. One of the popular notions is that ice melting is a natural consequence of particles getting into motion. For many students, ice has only fixed particles while there are loose and movable ones in water. One of the German students stated that "the fixed particles come into motion again (due to heat)", so the ice melts (from Grade 10). Other statements such as "The frozen ice particles hardly move (solid), but after taking in energy it moves and vibrates (liquid)" and "Particles in frozen state are bond to each other firmly and all stay in their own positions" are common among the students. Very straightforward is the statement from a German 10th–grader that "…the melted particles are no longer fixed, but rather slightly *fluid* and in motion" (emphasis added). A German 11th-grader illustrated his idea of fixed particles in ice and the freely moving ones in water as seen in Figure 8. A student of Grade 13, holding the same idea, added that the particles in ice possess no kinetic energy. All these examples show that there is a clear parallelism between ice/fixed particles and water/moving particles. Similar to the findings from their Taiwanese counterparts, the microscopic view is reflected macroscopically, and in this way what is seen (ice and water) and what is beyond (particle motion) correspond with each other.

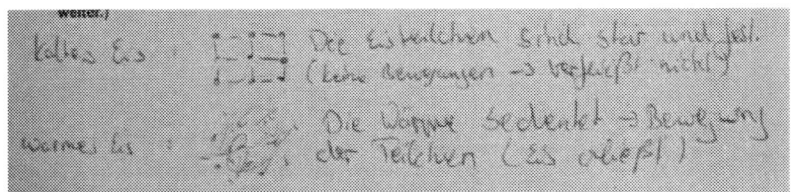

Figure 8. Particles are fixed in ice and freely moving in water, as illustrated by a German 11th-grader.

It is thus understandable that ice melting is explained as a result of particles becoming movable due to the higher temperature outside. As a German 13th-grader commented, "Outside is warmer, so the particles come into movement, and can no longer be held in structure". An alternative, but similar, argument is also put forward: "The static particles are replaced by moving ones", and consequently ice melts (German 10th-grader). For many students, ice becoming liquid and its particles getting in motion (or replaced by those in motion) are corresponding events. It seems that there is a direct linkage between the move of the particles and the move of liquid, the microscopic and macroscopic phenomena. The solid being fixed or the liquid moving freely among itself become manifestation of their particles respectively being fixed or movable.

Evidence on students' reflection of microscopic actions based on macroscopic appearance is further provided when some German students describe the particles in the melting process in terms of "expand" (*dehnen sich aus*). To take an example, a 10th–grader and a 11th-grader similarly commented that "the particles expand" when ice is heated and starts to melt. An illustration, as Figure 9, from another student vividly disclosed this idea, in which the particles in "cold" (such as ice) are concentrated and those in "warmth" (or heat; such as water) dispense.

Figure 9. Particles in cold (*Kälte*) and in warmth (*Wärme*), illustrated by a German 10th-grader.

Ice melting is sometimes explained by the German participants as a result of "friction" among particles. Several students seem to believe that ice melts under room temperature because its particles start to move (due to the warmth), "rub" each other and in turn generates heat, which causes the melting. As a German 10th-grader wrote, "The particles of the ice get into motion again, which they don't in a cooled condition, and rub each other. This gives rise to heat. Then ice melts". For the student, ice melting is actually caused by an "internal friction heat" which is triggered by the external warmth. A similar idea is put forward by another student: the particles in the warmer air rub the ice's "fixed atoms" and in turn "friction energy" is generated. Here the friction energy is till the cause, but generated between ice and air particles instead of between ice particles themselves.

Due to this energy the structure is loosened, which leads to the melting of ice, as added by a German 10th –grader. Similarly, another student commented that when the particles move, friction will be generated and in turn make the temperature higher. It should be also noted that ideas as such disclose a confusion of cause and result. The addition of heat as a cause of ice melting is replaced by an imaginary internal force called friction among particles as a result of particle motion. The idea is erroneous, yet intriguing. It should be noted that such idea reflects again students' misuse of macroscopic concept (friction) in describing and predicting the atomic phenomenon (particle behavior in the process of ice melting).

Several German students from Grade 11 specifically mentioned about the collapse of the structure while ice is melting, as one of them wrote, "the fixed framework (*Gerüst*) of the ice are destroyed, so that it becomes fluid." It is, however, worth noting that only a small number of students came to the point of thermal equilibrium and recognized that temperature difference and the direction of heat flow from higher to lower temperature are fundamental to the mechanism of this process. It is correct to say that ice starts to melt because heat is added. But a sufficient understanding for a secondary student would necessarily include the idea that the difference in temperature causes this heat transfer. It seems that the participant students are more aware of the "presence" of heat in the environment than the "transfer" of heat between the object and the environment in the process of ice melting. Their understanding of heat remains at a superficial level where its transitional feature is often dismissed.

DIFFERENCES BETWEEN PARTICLES IN SOLID AND IN LIQUID PHASE

The majority of the students held true that the state difference is associated with the motion of and the distance among particles. Some of them held correctly the idea that particles move faster, are less bound to and have more distance from one another in liquid than in solid phase. However, as already discussed, a considerable number of the students believed that ice have fixed particle while water have movable ones. It is relatively easy for students to come to the conclusion that water particles have greater distance among each other than those in ice. Based on such ideas, a number of the German students illustrated similar pictures of particles in water and in ice; examples can be seen in Figure 10, Figure 11 and Figure 12.

Figure 10. Particles in water and in ice (1), illustrated by a German 10th-grader.

Figure 11. Particles in ice and in water (2), illustrated by a German 10th-grader.

Figure 12. Particles in ice and in water (3), illustrated by a German 10th-grader.

It is also noticeable that students sometimes describe a microscopic event or state using macroscopic terms. It is, for example, commented by a 12th-grader that "The water bridge bonds are *melted*" (emphasis added). Other extracts referring to density also demonstrates this kind of confusion, such as "The particles of the frozen ice have a higher *density* (*Dichte*) than those in the unfrozen" (German 12th-grader) and "(particles absorb heat and obtain more kinetic energy) …the *density* is therefore reduced, which leads to ice becoming water" (German 13th-grader). It is held true by a number of students, in particular the Taiwanese, that particles in water have a higher "density" than those in ice. Several students believed that particles in ice are "cold" as opposed to the "warm" particles in water. This again provides evidence that the students do not distinguish the macro- and microscopic qualities.

Among the German sample, there seems to a correspondence between the school grade and the use of scientific terms. Students in the higher grade tend to use more scientific terms in responding to the questions than those in the lower grade. Common phrases such as heat (*Wärme*), temperature (*Temperatur*), heat transfer (*Wärmeübertragung*), heat conduction (*Wärmeleitung*) and particle motion (*Bewegung der Teilchen*) are used among all students, but more specific terms such as phase (*Aggregatzustand*), attraction (among particles) (*Anziehung*), kinetic energy (*Bewegungsenergie* or *kinetische Energie*), melting point (*Schmelzpunkt*), particle energy (*Teilchenenergie*), speed (*Geschwindigkeit*), heat conductivity (*Wärmeleit- fähigkeit*), particle bond (*Teilchenverband*), heat influence (*Wärmeeinfluss*), energy exchange (*Energieaustausch*), water bonding bridges (*Wasserstoffbrückenbindung*) and particle structure (*Teilchenstruktur*) appear more frequent, if not only, in the responses of the students from the higher grade. It seems that students of the higher grade use more "scientific language" to describe a similar idea than those of the lower grade. For example, it is more common among students in the higher grade to use "vibrate" (*schwingen*) to describe particle motions instead of "move" (*sich bewegen*). Their responses may not necessarily be more accurate, but tend to be more sophisticated than what the younger students commented.

It should be noted that thermal equilibrium seems to be a more familiar concept for the Taiwanese than for the German students in the study. The textbook analysis shows a correspondence to this result (to be discussed in the next section). The Taiwanese textbooks have a clear emphasis on the explanation of thermal equilibrium, in particular while temperature is introduced, whereas few notions on this concept are found in the German sample.

Chapter 5

THE SCIENCE OF HEAT IN TEXTBOOKS

This chapter discusses the analysis of science textbooks used in the targeted student group. The analysis is intended to reveal the presentations of the science of heat in the teaching/learning material and thereby to identify possible connections to students' ideas. It should be noted that textbooks in Taiwan are written according to more specific guidelines than those in Germany. That is to say, the structure and content are similar among the Taiwanese textbooks, while the German ones can vary significantly from one to another. There are also many more versions of textbooks available in Germany than in Taiwan, especially for each state in Germany has its own curricular guidelines. To correspond with the student sample, ten textbooks, among which are two teachers' handbooks, from Germany are selected for analysis due to their popularity in the state of Lower Saxony. For the part of Taiwan, ten commonly used textbooks - five student textbooks and five supplementary teachers' handbooks - are chosen. Textbooks which are used by the participant students are all included in this analysis. The analysis follows two principal questions: (1) What are the scientific meanings of the basic thermal concepts, especially heat, as presented in the textbooks? (2) What are the ways in which these concepts are introduced? Particular attention is given to the possible linkage of the textbook contents to the students' misconceptions as revealed in the investigation.

It should be noted that textbooks are used often as an important, but not only, guiding and reference resource in the both countries. Thus, these studied textbooks do not represent how the studied subject is actually taught, but rather the commonly accepted ideas about what and how it should be taught. In the following discussion, each textbook is represented by a three-letter code. To avoid commercial considerations, the textbooks are listed in the appendix without noting their codes.

THE DEFINITION OF HEAT

Several scientific definitions of heat are found in the textbooks under study (Table 18). Heat is most frequently explained as the energy transported from a hotter to a colder object. Other notions such as heat as a form of energy (being absorbed or released while the object is experiencing change in temperature or in state), as part of change in internal energy, as a kind of energy that can do work are also suggested. It seems that heat can be explained in many different ways, and, thus, may easily lead the learner to confusion. It is found that the analyzed textbooks' explanations on these definitions are often too brief and vague and their connections to other concepts, such as temperature and internal energy, are

Table 18. Definitions of heat provided by textbooks.

	German Textbook(s)	Taiwanese Textbook(s)
Heat is the energy transported from a hotter to a colder object.	DBI, MPI, PGI	OPI, OPL, HPI, HPL, DPI, DPL
Heat is the energy absorbed or released while the object is experiencing change in temperature of in phase.		SPI, SPL
Heat is a form of energy.		NPI, NPL
Heat is (part of) the change in internal energy	KPIII, KPII	
Heat is the energy being transferred from a hot to a cold body by means of irregular molecular motions.	DBI, KPI	
Heat is what is emitted from a hot source.	KPI	
Heat is a kind of energy that can do "work".	PGI	
Heat is entropy.	DKL	

rarely clarified although the text frequently, and inevitably, involves such connections. For example, one set of Taiwanese textbooks (including a textbook and its teacher's hand book) SPI and SPL state merely, though correctly, that heat is the energy absorbed or released while the object is experiencing change in temperature of in phase, without giving examples or making references to previously discussed concepts. Similarly another set of Textbooks, NPI and NPL, define heat as a form of energy without further clarifying its actual meanings in thermal phenomena. This draws a correspondence to the results of the investigation with students. The multiple, yet disconnected and sometimes even erroneous, meanings of heat as presented in the textbooks may help to explain why students describe heat in various, but confused, ways. It is especially worth noting that similar misconceptions to the student's are also revealed in textbooks.

HEAT IS THE ENERGY TRANSPORTED FROM A HOTTER TO A COLDER OBJECT

It is the contemporarily accepted scientific notion that heat is the energy which flows form a body with a higher temperature to another with a lower temperature. This notion is thus also mostly found in several textbooks under study. For instance, as stated in the DBI, a German textbook, "Heat is the energy being transferred due to the temperature difference. It always flows automatically from the hotter to the colder object" (p.230). In another German textbook, it is pronounced in a different way, but with the same meaning: „Let two object touch (or mix) each other. Their difference in temperature will cause a transfer of energy. In this case, it is heat being transferred" (PGI, p.246).

It is found that the Taiwanese textbooks seem to give more detailed explanation on this notion than the German sample. In the OPI, for example, the explanation starts with a contextual problem: If we pour cold water into hot pot, the temperature of the water will go up, whereas the pot will become colder. A question is then raised: What is actually happening during the time? At this point, the text goes on to the pre-scientific ideas and provides a short overview on the discovery of the current concept that heat is a kind of energy in transition. The explanation closes at the main argument that when two objects of different temperature come in contact, energy will go from the one with higher temperature to the other with lower temperature until they reach the same temperature, a state of thermal equilibrium. The energy that is transported during the time is called heat energy (OPI, Taiwan).

In general, the Taiwanese Textbooks seem to introduce this meaning in a better defined context. In the following is another exemplar text:

> heat as energy in transition can be explained in the context of two bodies in contact. When we let two objects of different temperature get in contact, the

hotter will become colder, while the colder becomes hotter. This will continue until the both have the same temperature. In this process, heat is being transferred from the hotter to the colder objects. (HPI, p.65)

The textbook HPI goes on to ask "What is the nature of heat? Is it material or energy?" and seeks to answer these questions through historical thinking from caloric to kinetic theory of heat, and concluded that through Joule's experiments the concept that heat is a form of energy is then established.

It is worth noting that one of German textbooks seeks to clarify this concept by categorizing heat as "energy in transition" as opposed to "energy in storage":

While the energy forms such as potential energy, kinetic energy and internal energy are used to describe an existing state, the terms heat and work refer to energy in transition (*Energieübertragung*)...Potential energy, kinetic energy and internal energy can be considered as energy in storage (*Speicherformen der Energie*). (KPI, p.163)

Similarly in one of the Taiwanese teachers' handbooks, it is stressed that heat as "energy in transition", distinguished from "energy in a body", should be pointed out. The text reads:

Heat is a kind of energy which is transferred between objects or between the object and its environment in the *heating process*. It is called heat energy. It should be noted that this concept only refers to energy in transition. Thus, it is correct to say a body receives or releases how much (quantity of) heat, but incorrect when we say a body has how much (quantity of) heat. (DPL, p.69, emphasis added)

Although this clarification should meet the need to minimize students' confusion which is reported in the last section, the use of the emphasized phrase "heating process" may contribute to another misconception which students often hold that heat transfer is associated with a hot source and heating (note: cooling also involves heat transfer). It should be noted that phrases used in several textbooks can be misleading as such, even though what they intended to explain is of importance.

HEAT IS (PART OF) THE CHANGE IN INTERNAL ENERGY

There seems to be a tendency that the German textbooks deal with heat after temperature and internal energy are introduced. Heat is thus often discussed in relation to internal energy; for example, as stated in one textbook, "internal energy of a body increased when heat is added" (KPIII, p.167). Another German textbook also pronounces this meaning, and gives the following explanation:

> Heat can be understood in terms of the change in internal energy... If work is also involved at the same time, it should be taken into consideration. Heat is then reduced to a part of the internal energy change... Heat is defined through the equation $Q = \Delta U - W$ [where Q is heat, ΔU is the change in internal energy and W is work]. (KPII, p.26)

This explanation includes the mathematical equation involving three important concepts, internal energy, work and heat, yet does not offer relevant examples, nor additional reminder on the meanings of these concepts.

HEAT IS THE ENERGY BEING TRANSFERRED FROM A HOT TO A COLD BODY BY MEANS OF IRREGULAR MOLECULAR MOTIONS

A German textbook explains heat simultaneously from both macro- and microscopic perspectives: "Energy, which is transferred from a hot to a cold body by means of irregular molecular collision, is what we call heat (*Wärme*)" (DBI, p.230) Such a explanation can be considered to be a basis of or support for students' mixed descriptions of macro- and microscopic phenomena. Its problem is the lack of a proper context where the two concepts, heat transfer from a hot to a cold body and heat transfer through molecular collision, can be integrated. Students may therefore miss the point that the former is in nature a macroscopic explanation and the latter a microscopic one.

HEAT IS A KIND OF ENERGY THAT CAN DO "WORK"

Heat is also discussed in the section "Machine, which works with heat (*Maschine, die mit Wärme arbeiten*)" in one of the German textbooks as the energy that can do work. It states that "the energy needed for doing work can be applied in the form of heat" (PGI, p.285). The concept of heat is explained in the text as a kind of energy that can do work. This is again a different, but also scientifically correct, meaning which is assigned to the concept of heat. It reconfirms the fact that heat can be explained in various ways and in relation to different concepts as accepted by science experts. In view of students' understanding of the heat concept, however, we should find this problematic as the multiple use of this term in the textbooks is manifest, but not clearly identified and differentiated. While the textbook offers the account that relates heat to work and energy, its underlying meaning (the *effect* of heat) essential to differentiate it from other meanings is nevertheless left unclarified.

HEAT IS ENTROPY

The analysis locates a crucial misconception in the German teacher's handbook DKL, where an entire chapter entitled "Entropy (*Entropie*)" is included. In this chapter the term "entropy" is used where heat and thermal energy should be placed. The term "entropy flow" (*Entropiestrom*) is consequently used referring to "heat flow." In the following are some exemplar statements:

> "The higher temperature a body has, the more entropy it contains." (p.120)
> "Entropy flows automatically from one place of higher temperature to another of lower temperature." (p.121)
> "A difference in temperature is the drive for entropy flow." (p.121)

This misconception is apparently passed on to the students in the study: As discussed previously, it is a uniquely supported notion among the German students that heat is the same thing as entropy.

HEAT HAS PRESENTED IN EVERYDAY AND SCIENTIFIC WORLD

In one of the German textbooks, a comparison of "everyday" and "scientific" explanations of heat is provided. It takes the example of a pot being heated on the stove, and describes fours steps of the heating process using the everyday and scientific language, as presented in Table 19 (original text and illustration in Figure 13).

Table 19. Heating process as described in everyday and scientific language.

	Everyday language	Scientific language
1.	The heat of the stove is too little.	The temperature is too low.
2.	The stove contains heat.	The hot stove has internal energy.
3.	The stove gives out heat.	The stove gives out heat through conduction.
4.	The stove radiates heat.	Releasing heat through radiation.

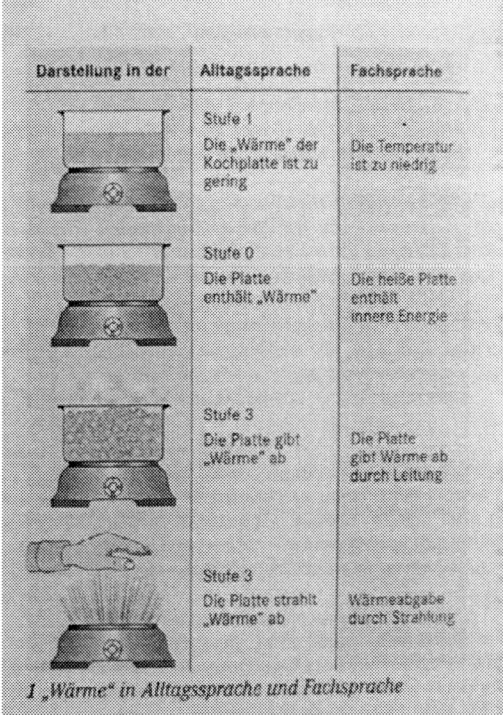

Figure 13. Heating process as described in everyday and scientific language. (KPI, p.162)

The text goes on to summarize several meanings of heat present in everyday context: "the state of being warm," "what is contained in a hot source, such as a stove," "what is transferred from a hot source to an object and makes it also hot," "what is radiated from a hot source, such as heater or oven" (KPI, p.162-163). The teachers' handbook produced by the same publishers, also calls for attention to the multiple use of the term "heat" in the everyday world. It points out that "the word 'heat' as used in everyday life can have many physical meanings" (KPL, p.65). These meanings are partially unique in its language context for the scientific term of heat is an identical word to warmth (*Wärme*) in German. As it contends, heat can mean:

Temperature: "A summer day – up to 30 degrees of warmth" (Ein Sommertag – bis zu 30 Grad Wärme),
Heat Energy : "The oven radiates heat." (Die Ofen straht Wärme ab.),
Heat Substance, Internal Energy: "Use the remaining heat of the stove while cooking." (Nutzen Sie die Restwärme der Herdplatte beim Kochen.),
Enthalpy (Heat Content): "Evaporating water gives heat to the environment." (Verdampfendes Wasser gibt Wärme an die Umgebung ab.), and
Entropy: "Through friction heat will be generated." (*Bei der Reibung wird Wärme erzeugt.*) (KPL, p.65).

It should be noted that this summary conveys some erroneous and misleading message, especially at the last two points, and should be reexamined. Moreover, summaries as provided by these two textbooks show a correspondence to what students hold as true in regard with the concept of heat. As discussed in the previous section, students simultaneously discuss heat as something hot or warm, as an internal quality of an object, as energy released or radiated by a hot object or system, etc. These ideas appear to be a mixture of the concepts of heat, temperature, hotness, internal energy and radiant heat, etc., which seems to have its root in everyday language. It should be noted, however, that a summary of everyday meanings of heat alone may not suffice to refrain students' misconceptions (especially evident in view of students' understanding of this subject). What is needed is a clarification of where and how a particular meaning is given to the concept of heat, and, in particular, of underlying principles which distinguishes these meanings.

HEAT AND INTERNAL ENERGY

In general, two explanations of internal energy (or thermal energy) are found in the textbooks: first, the energy which a body contains and is proportional to its temperature; second, the sum of potential and kinetic energy of the particles in a body. Several textbooks have commented on the importance of and the ways of distinguishing heat from internal energy. In one of the German textbook, for example, it is stressed that heat should not be confused with internal energy and the following explanation is given:

A body with higher temperature has more internal energy. If it gives out some of the energy, then this flow-out energy portion is called heat. When it arrives in the cold object, it is no longer heat, what is being raised is the internal energy and temperature. (DBI, Germany, p.230)

Especially worth noting is the use of some metaphors to distinguish internal energy and heat: for example, in the German textbook KPI, the difference between heat and internal energy is compared to "work done in lifting" (*Hubarbeit*) and potential energy (*Lageenergie*); more intriguing is the use of "bank transfer" as a metaphor in the German textbook DBI:

The money in a bank account is part of your property. You may transfer some of it, a so-called transfer amount while the money is on the way, to somebody. When it arrives at the recipient, it becomes his property. Energy can be explained in a similar way: what the body has is internal energy, and what is *on the way* is heat or work. (p.230)

Another German textbook, MPI, after defining heat energy as the energy being transferred, due to the difference in temperature, from one body to another, goes on to state that we cannot speak of "stored" heat in physical sense, because no matter mechanical energy or heat energy, once absorbed, they no longer differ from each other in nature. They all become part of the thermal energy in the particles.

It is also stated in two German textbooks that the internal energy can be changed through work or heat. In the case of heating, for example, energy can be added either through work or as heat. This energy is then stored in the body - more precisely, in the movement and the structuring of the particles – and is what we call internal energy.

HEAT AND TEMPERATURE

Although research has showed much evidence on students' problems of confusing heat and temperature, textbooks do not seem to react accordingly. Few efforts are made in the textbooks under study to insist students in distinguishing and relating the two concepts. Although a Taiwanese textbook seeks to clarify heat and temperature at the same time, it remains as a mere statement rather than a contextual or reflective argument:

Temperature is different from the quantity of heat. Temperature is to denote the degree of hotness and cold of an object, whereas (the quantity of) heat is the energy being absorbed or released while the object is experiencing change in temperature or change of state. (SPI, p.51)

Temperature as a basic concept in thermodynamics is of much weight in the textbook contents. Its several meanings are discovered and discussed in the following.

TEMPERATURE AS AN OBJECTIVE MEASURE OF COLD AND HOTNESS

As clarified in a German textbook, "vague descriptions such as "cold" and "hot" are specified [through temperature]... by exact numerical values with unity" (DBI, p.221). This idea seems to gain more attention from the Taiwanese textbooks while it is more frequently mentioned and its descriptions seem more telling. The exemplar text in the following explains the need for an objective measure instead relying on the sensorial feelings to indicate how warm or cold a body is:

Put your hands into cold and hot water separately, then together into warm water. The hand originally in the cold water will feel warm, whereas the other hand will feel cold. Thus, to judge the degree of cold and hotness of a body should not depend on sensorial feelings, but rather an objective measure. *This measure, which quantifies the cold and hotness of a body, is called temperature.* The device to gauge temperature is called thermometer." (NPI, p.80, emphasis added)

Similarly, it is argued in another textbook that sensorial feelings can not be quantified and are confined by our physical tolerance, and therefore it is necessary to have a measure such as temperature. The extract reads:

We can feel hot or cold... scorching, hot, warm, cool, cold, freezing, etc. reflects what we sense on an object in terms of cold or hotness. This is where the concept of temperature arises. However, simply replying on our sensorial perception to judge a body's temperature is neither objective nor scientific. Firstly, *we cannot quantify our feelings.* Secondly, *the range of hotness or cold we can test through our sensation is limited.* That's why we need to measure temperature by means of an objective method." (HPI, p.62)

TEMPERATURE AS THE QUALITY WHICH IS THE SAME IN DEGREE AMONG OBJECTS IN THERMAL EQUILIBRIUM

As explained by several textbooks, when two bodies with different temperature get into contact, the hotter will give out heat to the colder, until the both reach the same temperature – a state of thermal equilibrium. Thus temperature is defined as the common physical property shared by all objects in thermal equilibrium. In the German textbook MPI, for example, it is stated that "All bodies which are in the state of thermal equilibrium (*im thermischen Gleichgewicht*) have a common physical property which is called *temperature*." (p.146)

In the German textbook DBI, the concept of thermal equilibrium is explicitly explained. It first contends that "A plate which lies in the blazing sunlight receives the energy...in every second. However, its temperature will rise only up to a certain final temperature" (DBII, p.166). It goes on to use a water model as analogy (see Figure 14), stating that when the water which flows in and out the vessel has the same speed, there is a balance, where the water level is constant. The plate in the sunlight, similarly, reaches a final temperature when it releases as much heat as it absorbs from the sun. Before this, the temperature of the plate is lower, and more heat is being absorbed than being released. While the heat supply does not cease, the temperature rise does not stop until the "balance temperature" (*Gleichgewichtstemperatur*) is reached. This final state is termed thermal equilibrium (DBII).

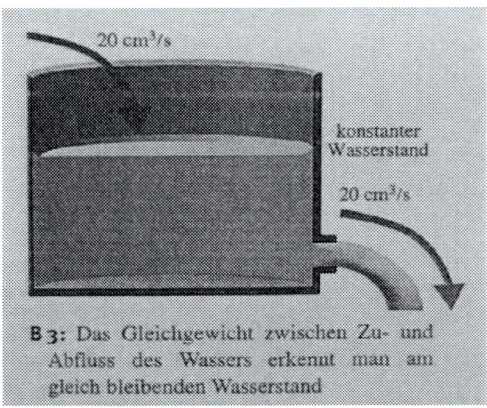

Figure 14. Thermal equilibrium analogous to water balance in a vessel. (DBII, p.167)

This kind of contextual explanation is however exceptional among the German textbooks. The others, despite mentioning it, do not seem to put adequate attention to explaining the concept of thermal equilibrium. Also, it occurs in one of the German textbooks that the concept of measuring temperature is incorrectly described as based on the fact that the temperature of the object is "adopted" by the thermometer (*dann nimmt dieser nach einiger Zeit die Temperatur des Körpers an...* (MPI, p.146)). This can lead to the misconception of thermal equilibrium as a result of one becoming the same as the other, which is present among students as the study reveals.

In contrast, all the Taiwanese textbooks give explanations to thermal equilibrium along with temperature, revealing an inseparable relation of the two concepts. In the following is an exemplar extract:

> If we let two objects of different temperature come into contact, they will reach the same degree of hotness and cold, a state of thermal equilibrium, after a while. Whatever quantity, shape, size and material, two objects in the state of thermal equilibrium have the same temperature. In other words, two objects with the same temperature are in thermal equilibrium with each other... Therefore, temperature is a physical property of a system that underlies the common notions of hotness and cold. (SPI, p.51)

The idea that two objects in contact will eventually come to a state of thermal equilibrium, where they share the same temperature, is further connected to the principle of temperature measure, as the above-mentioned textbook SPI cited that "*This concept that two objects in contact will eventually have the same temperature is a basis of temperature measure*" (SPI, p.51, emphasis added). This linkage among thermal equilibrium and temperature is explicated in two another Taiwanese textbook. They point out that the object and the thermometer in contact will come to thermal equilibrium and have the same temperature, so that what the thermometer reads is the temperature of the object, as an extract reads:

> When two bodies of different degree of hotness and cold get in contact, they will, after a long time, reach the same degree of hotness and cold. This final state is called thermal equilibrium. When two bodies are in thermal equilibrium, they have the same temperature. Thus, we can measure a body's temperature by means of thermometer which reaches thermal equilibrium with the body. (DPI, p.61)

As a coherent approach, it is emphasized in the Taiwanese teachers' handbook SPL to *introduce temperature through the concept of thermal equilibrium.* More explicit is the teaching/learning principle proposed by the another Taiwanese teacher's handbook DPL, where it contends that thermal equilibrium is a very basic concept in thermodynamics and can be used to define another basic concept, temperature.

It should be noted, however, explanations detached from context is no rare among textbooks. This is especially concerning because, as the investigation reveals, many students do not apply the concept of thermal equilibrium to familiar contextual problems despite stating the meaning correctly. It seems that the textbooks do not go beyond the descriptions of the concept to provide a range of phenomena for application. Without repetitively applying the concept to various phenomena students would have difficulties to catch the underlying principle and thus fail to reach a real understanding of the concept.

TEMPERATURE AS A RESULT OF MOLECULAR MOTION

The concept of temperature is also explained frequently in terms of molecular motion. In explaining the behavior of gas, it is often stated that the higher temperature a body has, the faster its molecules move. For example, as explained in one of the German textbooks, "the gas particles with rising temperature move quicker on average" and "[d]uring their quick move the gas particles also collide with one another. The result is a completely irregular movement of all particles" (DBI, p.228). In a Taiwanese textbook, it is stated that "Temperature reflects the degree of vigorousness of the particle motion" (OPI, p.88). As found in most of the textbooks, temperature is defined in the molecular level as "an indicator of the average speed of the object's particles" (German textbook PGI, p.242), or as "a measure of the average kinetic energy of the object's particles" (German textbook LKII, p.216). There are, however, few explanations found in attempt to differentiate or connect these two meanings.

A problem is especially identified in view of students' understanding of the temperature concept: While explanations may be correctly communicated, the perspectives from which they are given are often not specified. What may be difficult for students to grasp is such a change of explanation from temperature as a measure of hotness and cold to temperature as an indicator of particles' speed for they do not appear connected in their verbal expressions. To really understand it would require a clarification on the perspectives from which these explanations are generated.

PARTICLE MODEL IN THE CASE OF PHASE TRANSITION

Using particle model to explain phase transition is found in virtually all textbooks under study. Most frequently discussed is the melting process. As explained in the Taiwanese textbook OPI, when a solid being heated and its temperature rising, its particles move with a greater speed and amplitude, distances among particles increase, and the bonding among particles weakens. When the temperature is high enough, the particles move so fast that the attraction among particles does not suffice to keep them in position, they become looser and the solid becomes liquid. This change of state (phase) is called melting.

In order to clarify this idea, particle model is thus explained. As shown in Figure 15, the above-mentioned textbook illustrates particle model of solids, liquids and gases, giving the following summary:

In solids, due to the attraction among them, particles are kept in their own position and only move slightly;

In liquids, due to the higher speed of the particles, their bonding becomes weaker, thus they are more apart from one another and move faster;

In gases, particles move even faster, and can become totally free and rid of attraction.

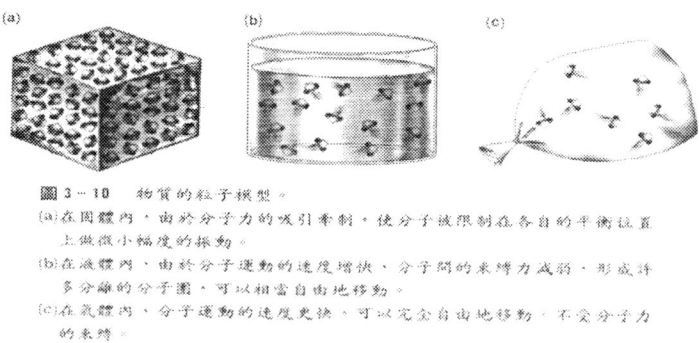

Figure 15. Particle model of three phases. (OPI, p.89)

Melting occurs, according to most of the textbooks, while heat is added and temperature is raised to the point where particles in the solid start to vibrate in a greater distance and no longer stay in the tight structure (Figure 16). It is sometimes pointed out that external energy (heat) is needed for the transition as such in order to increase particles' potential energy and to oppose the "attraction force" among particles. However, while several concepts are involved in explaining particle model, these concepts may not be fully understood by the students, and thus can lead to further misconceptions of particle model. While mentioning structure of particles, vibration, potential energy, attraction force, etc. there is little attempt observed to re-clarify or connect them into a more coherent piece of explanation, nor to entertain them using contextual problems. It seems that more emphasis is laid in the textbooks on demonstrating the knowledge itself than comparing and relating the knowledge. Having to keep up with the learning load, it is understandable that students merely memorize what is taught without really understanding it.

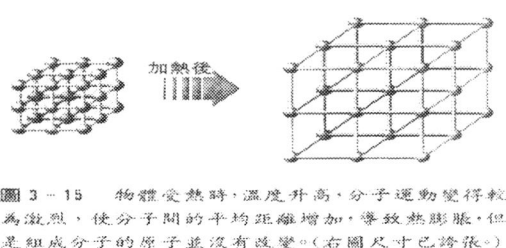

Figure 16. Particles are more distant from each other after heated. (OPI, p.93)

HEAT TRANSFER

Heat transfer is one of the core concepts elaborated in all the studied textbooks. Its direction, types and mechanism at the atomic level are discussed. It is apparent through the analysis that the secondary science of heat lays much emphasis on microscopic explanations of the thermal phenomena.

DIRECTION OF HEAT TRANSFER

In discussing the direction of heat transfer, the German textbook KPI cited that heat cannot automatically go from a body at a lower temperature to anther at a higher temperature; the natural direction of heat transfer is from the hotter to the colder. The concept that heat released by the hotter object releases will equal that received by the colder is then communicated along with its mathematical application. The German textbook DBI, for example, cited that upon mixing the hot and cold water, "the hot gives out part of its internal energy, and its temperature decreases... In such a mixing case the absorbed energy is always the same with the released." The formula $W_{auf} = W_{ab}$ (where *auf* indicates "received" and *ab* "released") follows (DBI, p.231).

It should be, however, noted that some textbooks do not specify the direction of heat transfer while discussing the mechanism of heat transfer, but instead stress the difference in temperature as the drive for heat transfer. This could be a weakness of the teaching strategy where the principles of the heat transfer are not clearly stated and linked.

TYPES OF HEAT TRANSFER

Most of the Taiwanese textbooks categorize heat transfer into three types: conduction, convection and radiation. These types are distinguished based on the ways in which heat is transported, especially from a microscopic view. As an example, the NPI describes these types as follows:

Conduction: by means of collisions among particles, and even electrons.
Convection: by means of the move of the particles in a body.
Radiation: without medium, in a form of electromagnetic wave.

In contrast, the German textbooks emphasized these types of heat transfer based on the criterion of whether or not, and how, a medium plays a role. As summarized in the DBI, the three types of heat transfer are defined as such:

Conduction: Heat moves "through" (*durch*) the material.
Convection: Heat moves "with" (*mit*) the material.
Radiation: Heat moves "without" (*ohne*) material. (p.244-245)

It is worth noting that several German textbooks exclude "convection" from the categories. A confusing statement is offered by the KPI that, in the case of convection, it is not heat, but internal energy, which is transported, thus convection is a type of "energy transfer" instead of "heat transfer". Other textbooks do not explain as such, yet similarly discuss heat transfer exclusive of convection. For example, an extract reads "We refer heat to a form of energy, which is transferred through conduction or radiation" (KPI, p.163); another excerpt also discusses the types of heat transfer exclusive of convection:

Heat energy is the energy, which moves automatically from a hotter to a colder body. This transfer can occur through material between objects, in the form of so-called heat conduction, or through heat radiation... without any material contact. (MPI, p.157)

It seems that heat transfer is categorized by merely based on whether or not a medium is present, and therefore convection is seen as a kind of conduction. In one textbook, for example, it states that heat transfer which occurs "in a resting object", "by fluid" or "by gas" (with medium) can be denoted as "conduction" (KPI). Nevertheless, conduction is more commonly specified as heat transfer in and between solids. It becomes evident at this point that discrepancies are present among textbooks in explaining the types (or means) of heat transfer.

Some attention should be also paid to the meaning of heat radiation as students appear to have some difficulty with this concept. It is pointed out by a Taiwanese textbook that heat radiation is not a one-direction transport, which means, a body emits heat, but meanwhile attracts heat (DPL). It is also noted by a German textbook that all objects "send out radiation due to their molecular movement", and the intensity of the radiation is dependent on the speed of the movement, and consequently on temperature (while temperature represents the molecular speed) (DBII). Given that participant students often confuse heat transfer as a result of temperature difference and heat being constantly released (radiated) from all objects, heat radiation can appear to conflict the concept of heat transfer, and thus requires a further explanation. Some effort should be made to clarify the idea that while heat radiation does happen upon all objects (above absolute zero) difference in temperature leads to a difference in the totals of heat flow between the two objects in contact.

The majority of the studied textbooks highlight heat conduction while explaining heat transfer. Heat conduction occurs, as the Taiwanese textbook HPI stated, when heat is transferred with the body from the part of higher temperature to another of lower temperature. Caution should be, however, called to avoid equalizing heat transfer and heat conduction. The three types of heat transfer should be clarified in such a way that their differences and connecting points are brought into light.

MICROSCOPIC MEANING OF HEAT TRANSFER

As mentioned in the previous subsection, heat transfer as explained using particle model is a familiar topic among the textbooks. Among others, heat conduction is most frequently discussed. It is argued that heat transfer can be best explained from the microscopic perspective: The higher-speed particles in the hotter body collide with the lower-speed ones in the colder body when they come into contact, and give over part of its kinetic energy (KPII). An illustration, as shown in Figure 17, is further provided to visualize this process. It is added in another textbook that heat transfer can also mean "the transfer of kinetic energy" at the atomic level (KPIII, p.169).

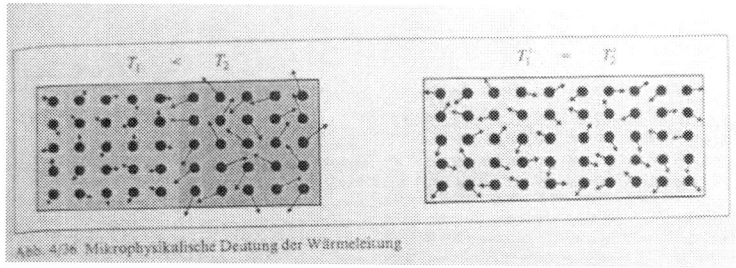

Figure 17. Heat conduction through particle collision. (KPII, p.42)

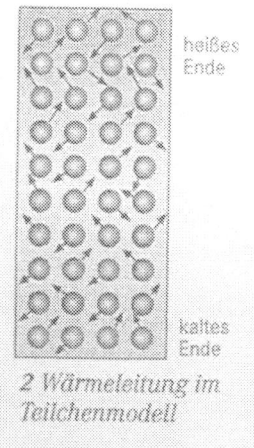

Figure 18. Heat conduction in particle model.(KPI, p.166)

Similar explanations are offered in another German textbook KPI: the fast particles (in the hotter object) hit the slow particles (in the colder object) and thereby turn over a part of their energy; this proceeds further to other particles until all particles reach the same speed, that is to say, the two objects reach the same temperature. It is also noted that heat conduction within one single object shares the same mechanism. Although an illustration is also provided (see Figure

18), it does not seem to convey the main principle while actions of particles at "the hot end" (on the top) and "at the cold end" (on the bottom) do not appear to differ from each other.

To explain the process of heat conduction within one body, another German textbook offers a somewhat different and confusing account: heating a cold object on its one end, the particles there will then vibrate with higher amplitude and frequency; the "force" between particles "transfer the strong vibration to their neighbors" and thereby all particles of the object (LKII, p.216). The illustration (see Figure 19) it offers to explain the generation of friction heat also appears problematic, for no sufficient clarification is provided to explicate the particle behaviors and the role of "friction" in this process. To reflect on students' understanding of heat, the found misconception of friction as a means by which heat is transferred seems to find its root in the textbook contents as such.

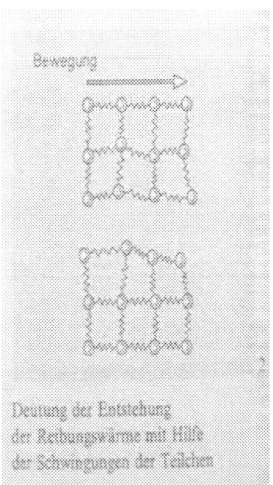

Figure 19. Heat conduction; its caption reads "The analysis of the generation of friction heat by means of the vibration of particles" (LKII, p.216).

It is also worth noting that textbooks sometimes do not specify the context while explaining heat transfer. For example, they may keep referring to heat transfer in general while they in fact describe heat transfer as it occurs inside a solid body, which means, heat conduction. The fact that concepts are sometimes not located in an appropriate context may contribute to students' confusion of mechanism of heat transfer and the differentiation of heat transfer, in general, resulted from a temperature difference, and heat radiation, specifically, independent of temperature.

Chapter 6

ISSUES REGARDING TEACHING METHODS AND SEQUENCES

THE USE OF HISTORICAL MATERIAL

Historical accounts are offered in several textbooks, concerning mostly the nature of heat and partly thermometer. The evolution of the ideas about heat from caloric theory through kinetic theory to heat as energy is frequently mentioned. A list of topics included in the textbooks can be summarized:

Theory change regarding the nature of heat: from caloric theory through kinetic theory to the concept of heat as energy. (DPL, PGI, KPI, OPI, NPI, SPI, DPI,)
    Beginning of the modern heat concept: Steam engine. (PGI, KPI)
    Temperature and heat as difficult concepts before the eighteenth century. (KPL, NPI)
    Temperature and heat distinguished by Black. (OPI)
    Thermometer: Fahrenheit, Celsius and Kelvin. (OPI)

While the evolution of the ideas about heat is the most popular historical topic, virtually no textbook, however, discusses the background in which and factors by which the transitions took place. Although one German textbook (PGI) provides several anecdotes about scientists such as Robert Mayer and James P. Joule, the intention seems to be more motivational than conceptual. There is hardly a connection to be located between the anecdotes despite the shared topic and scientific community. What the textbooks reveals is merely historical "facts" such as what was once held true in the past and who has suggested what, while the more telling and instructive information should be why and how it was once thought that way and was later defied. In the Taiwanese teachers' handbook KPL, it is mentioned that as a consequence of the previous scientists' attempt to clarify the concept of heat, the microscopic theory of heat was being developed, which attributed to the termination of the caloric theory. This is the closest in all textbooks to an argument linking two theories (caloric and kinetic theories). It remains, however, an abrupt conclusion which fails to bring into light the ways in which a theory evolves or becomes replaced by another.

Much research has been conducted to investigate and discuss the linkage between historical development and students' learning of science. Yet, textbooks do not seem to pay comparable attention to this issue. A relevant passage is found in one Taiwanese teachers' handbook, stating that historical development shows certain problems in this domain of knowledge which challenged the early scientists and are still present among today's learners (KPL). Yet, no further clarification is given.

CONFUSING TERMS AND ILLUSTRATIONS

Some problems with terms and illustrations used in the textbooks are additionally revealed, especially in view of students' misconceptions. Sometimes, more than one term is used to refer to the same concept, which can lead to confusions and misunderstanding. It also occurs that several scientific terms are used at the same time to describe or explain a concept or a phenomenon, and in turn complicate and entangle the presentation. As noted previously, illustrations in the textbooks may be problematic, albeit they are intended to assist the learner in constructing mental picture of abstract ideas.

The term "heat quantity" (*Wärmemenge* in German; *Re-Lian* in Chinese) is frequently used in place of heat in both German and Taiwanese textbooks. Moreover, heat, heat energy and heat quantity are often used without differentiation.

Several Taiwanese textbooks use the word "*Re-nen*" (heat energy) to refer to thermal energy without a further clarification, which may easily give rise to misconceptions. For example, in the section of "Heat Transport", the textbook NPI contends that "This movement [of particles] is manifest of heat energy (*Re-nen*), and temperature is the measure of the amount of heat energy (*Re-nen*)" (p.87), while thermal energy is meant here.

The meaning of thermal energy is not agreed upon by all textbooks. Some suggest it as potential energy of the particles, while the others define thermal energy as the sum of potential and kinetic energy of the particles. The textbooks do not provide contextual information so that these two meanings can be reconciled.

Some textbooks describe and explain a concept or theory using many different, and often difficult, scientific terms, and thus complicate and entangle their presentations. In explaining the main principles of mechanical heat theory, for instance, a number of terms are used in a German textbook within one page, including Newtonian mechanics (*Newtonsche Mechanik*), mechanical energy form (*mechanische Energieformen*), kinetic energy (*kinetische Energie*), potential energy (*potentielle Energie*), internal energy (*innere Energie*), atomistic structure (atomistische Struktur), intermolecular force (*zwischenmolekulare Kräfte*), restoring force (*Rückstellkraft*), attraction force (*Anziehungskraft*), balance position (Gleichgewichtslage), static statements (statische Aussagen), static physics (*statischer Physik*), etc. (KPIII, p. 159).

Two German words, *Lageenergie* and *potentielle Energie* are used simultaneously to refer to potential energy. In lack of clarification, this may lead to some confusion as well.

The German word, *Temperaturstrahlung* (which literally means temperature radiation), is sometimes used to refer to heat radiation (*Wärmestrahlung*). This may cause and support students' misconceptions of heat as a similar notion to temperature.

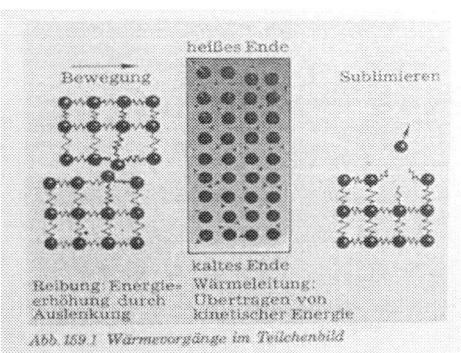

Figure 20. Heat transfer in the molecular world, in the case of (from left) friction, heating and sublimation. (KPIII, p.159)

Some illustrations provided in the textbooks seem to be as difficult to comprehend as the verbal explanations, and meanwhile fail to demonstrate the principal features they should be intended to reveal. As an example, the illustration (see Figure 20) in one of the German textbooks seeks to explain heat transfer in the molecular world, in the case of (from left to right) friction, heating on one end and sublimation, yet does not seem to explicate the main principle of heat transfer in general, nor the special features in these cases, as seen from an microscopic perspective.

Chapter 7

CONCLUSION

As the study reveals, despite years of science lessons, many secondary students fail to really understand basic thermal concepts and present the learnt knowledge in a confusing or/and disintegrated manner. Heat, as a core concept in the thermal science, is understood at a superficial level, as students may state correctly its nature as a kind of energy, yet turn to the idea of material-like heat in solving some problems. This indicates that the concept of the material-like heat still finds its position, and, moreover, forms part of a "hard core," in the student's conceptual framework (as previously argued by Niaz (2006), based on his study with science major freshman students). Most evident is that students describe and explain the concept of heat in association with other thermal concepts and properties, such as hotness, temperature, internal energy and heat source, yet the differentiation and relation among these ideas are frequently missing or misplaced. While this core concept is not adequately understood, the conceptualization of related thermal principles and other thermal concepts is thus problematic. For example, the popular misconception of heat as "energy in a body" understandably leads to the idea that heat transfer is resulted from the difference of heat amount between two objects.

It should be noted that the problems of students' understanding of this subject seem to correspond with the subject matter dealt with in the textbooks. As an example, thermal equilibrium seems to be a more familiar concept for the Taiwanese than for the German students in the study, and the textbook analysis shows a correspondence to this result. The Taiwanese textbooks have a clear emphasis on the explanation of thermal equilibrium, in particular while temperature is introduced, whereas few notions on this concept are found in the German sample. A warning result is that erroneous clarifications are found here and there, such as using "entropy" in place of "heat," and explaining heat transfer in a way that heat radiation is excluded and in turn becomes a conflicting concept.

More obviously, textbooks often explain concepts without a clear reference to the context they are based and the perspective from which they are viewed. Students, for example, frequently describe the molecular properties and actions based on what is observed, such as believing that particles in solids are "fixed" while those in liquids flow "freely." This may be resulted from the fact that textbooks provide few accounts on specifying the macroscopic or microscopic context in which a concept, an event, a property is being discussed; What is also missing is the accounts of comparing and integrating the descriptions from these two views. In the following some remarks based on such results will be discussed with a view towards a reconstruction of the subject matter for better teaching and learning.

The Teaching of the Thermal Science Should Be Based on a Sufficient Understanding of the Subject Itself and Students' Learning of this Subject

Although textbooks do not fully present how the subject is taught in reality, they communicate the commonly accepted ideas of what should be taught and how. The problems of lacking "bridging" explanations among concepts, phenomena, contexts and perspectives as revealed in both textbooks and students indicate the importance of reconstructing the subject matter to be a more precise, unified and integral whole and to assist students in distinguishing and relating the learnt knowledge for a better understanding.

More Efforts Should be Made to Assist Students in Differentiating Heat, Temperature and Internal Energy

It seems that to differentiate heat, temperature and internal energy (or thermal energy) is an important step for constructing an appropriate conceptual map to organize these and further related concepts. To do this, the underlying principles of the basic concepts should be clarified. Temperature and internal energy, for example, represents respectively the "intensity" and "totality" of the warmth of a body, and heat and internal energy are indeed the same in nature, yet the former is "in transition," while the later is "in storage." These essential points should be explained in detail by contrasting the microscopic and macroscopic perspectives and through contextual problems.

THE DIVERSE MEANINGS OF HEAT SHOULD BE CLARIFIED IN REFERENCE TO THEIR ONTOLOGICAL COMMITMENTS OR CONTEXTUAL BACKGROUND

It is evident that students often fail to bring together the diverse meanings of heat in a consistent manner. Their discussion of these meanings indicates, similarly, their lack of understanding of the background knowledge to which these meanings are attached. The diverse meanings of heat should be thus clarified in such a way that their ontological commitments or contextual background become apparent and can further help the student to differentiate these meanings. For example, while heat is explained as the energy going from the hotter to the colder in one place and as the energy which can do work in another, it should be made clear that the former is intrinsically a discussion on "the nature of heat" and the latter on "the effect of heat." Similarly, the nature of heat are typically introduced in two ways: a form of energy in transition and a manifest of particle motions, yet without clarifying the macroscopic and microscopic context which they are respectively embedded, the student would have difficulties (as the study shows) in bringing these two notions together and applying them where is appropriate.

HISTORICAL MATERIAL CAN BE EMPLOYED TO CONVEY UNDERLYING MEANINGS OF THE HEAT CONCEPT

The science of heat should be reconstructed in such a way that the underlying meanings of the concepts become visible so that these concepts can be distinguished and related. Historical analysis of the scientific knowledge in question is an important means to tackle this reconstruction, especially for we may convey elements and contexts of foundation and development of theories which are not novel in students' conceptual framework, e.g., relating heat to hotness, to substance/fluid, to motion, or to temperature. Particularly instructional meaningful is to clarify the process of theory change in such a way that the features which are important for discerning and relating one theory/concept to another become apparent. For example, the arguments and experiments of friction leads to the decay of caloric theory, and when heat as motion comes to discussion, the question about what is actually moving leads to the microscopic level of explanations of heat. When such sequences and contexts are explicated, students may see the underlying meanings of the concepts in question and be able to construct a conceptual map to accommodate and relate these concepts. As argued by Ling et al.(2006), the learning outcome depends on "whether students are provided with the learning experience that was supposed to help them discern the critical aspects".

EXPLAINING A CLUSTER OF CONCEPTS THROUGH A RANGE OF CONTEXTUAL PROBLEMS SHOULD SUPPORT STUDENTS IN CONSTRUCTING THEIR CONCEPTUAL MAPS

Conventionally the subject matter is introduced from one concept to another, and examples are made to explain a particular concept. What should be more instructionally fruitful is to explain a cluster of concepts through a range of phenomena; which means, a concept is repeatedly discussed in relation to other concepts in various settings. In this way, attributes and relations of the concepts can be more readily recognized and further organized by the learner. In the following some examples based on the contextual problems involved in the investigation with students are discussed:

Phenomenon 1: Ice being placed in the room and beginning to melt. This would have to include the discussion of (1) the action that ice absorbs heat, (2) the mechanism of this action (heat flows from the hotter to the colder; thermal equilibrium), which could also bring about the discussion of how to determine which is hotter/colder, (3) the idea of "phase" (the visible and invisible (molecular) features), (4) "phase transition" as result of heat transfer, (5) this effect as opposed to "temperature change" as a result of heat transfer, (6) what is heat as viewed through this phenomenon.

Phenomenon 2: Metal feeling colder than wood. In explaining the reason for this phenomenon, the teacher can similarly approach some fundamental thermal topics such as (1) the feel of hot or cold as a result of heat transfer, (2) the mechanism of heat transfer, (3) the differentiation of the "feel of hot or cold" (sensorial measure) and the "actual degree of hotness or cold" (temperature), (4) what is heat as viewed through this phenomenon.

Phenomenon 3: A cup of cold water and a cup of hat water being mixed. Should the student observe the two cups of water for a while before the mixing, the phenomenon of the cold becoming slowly warmer and the hot becoming colder can help to embark a discussion of some basic thermal concepts and principles such as (1) the mechanism of heat transfer, (2) the concept of temperature (3) the concept of heat. Upon mixing the two cups of water, these topics can be reviewed through a new observation, and thereby students' understanding can be deepened.

Through familiar thermal phenomena as such, questions can be raised along with observations and experimentations which repetitively draw near the discussion of the core concepts and principles of this subject.

DISCREPANT EVENTS CAN BE USED TO ASSIST STUDENTS IN DISCOVERING THE WEAKNESS OF THEIR CONCEPTIONS

The common-sense idea that things which feel hotter have higher temperature leads to students' confusion in the notion of "something feels hot" and "it is really hot". As an important point for understanding the concept of temperature, the sensorial feel of hotness and cold should be discussed and concluded with the need for a scientific measure. To expose the problem of believing that what something feels is what something really is, a discrepant event can be helpful. For example, let the student put one hand into cold water, and the other into hot water (not too hot, of course), it can be easily concluded that the former is colder than the latter. Yet, let the both hands put into the same water immediately afterwards, the two hands will feel very differently. This explains why the sensorial feel does not tell us really the object's state of hotness and cold. Moreover, this also conveys the fact that the sensorial measure is subject to the physical tolerance, and can only apply to a limited range of hotness and cold. Using a discrepant event as such can therefore help students to realize the weakness of the common-sense idea and to grasp the meaning of temperature.

INSTRUCTIONAL EFFORTS SHOULD BE MADE TO ENHANCE AND MAINTAIN THE CONSISTENCY AND COHERENCE OF THE SCIENTIFIC LANGUAGE USED IN THE CLASSROOM

As the study indicates, despite a familiarity with a number of scientific terms, students do not really understand their meanings and in turn often misuse them or confuse one with another. The textbook analysis reveals a correspondent presentation of the scientific terms. The scientific language used in the teaching and learning material appears to be inconsistent and sometimes even erroneous. Particularly worth noting is the fact that the scientific language is bound to the cultural context where it is used. Some terms are thus unique in a given cultural context and associated with its everyday language. For example, the commonly used German term, *Wärmezustand*, refers to the state of hotness or warmth, yet, can be understood as "the state of heat" while "heat" as a scientific term shares the same word with "warmth". This may contribute to the confusion between heat as a transitional energy and hotness as a state of being hot. Take another example from the Taiwanese sample, the word for "heat energy" does not differ from the word for "thermal energy," and the word for "heat" is exactly the same with "hotness." It is thus little wonder that students possess misconceptions based on such parallelism. Therefore, the language used for instruction should be carefully revised and used in caution. Which term for which meaning should be clearly indicated, and special attention should be given to those phrases and terms which are commonly used, yet based on everyday language which implies some conflict with the scientific concepts to be taught.

REFERENCES

Albert, E. (1978). Development of the concept of heat in children. *Science Education, 62*, 389.

Appleton, K. (1985). Children's ideas about temperature. *Research in Science Education, 15*(1), 122-126.

Bruner, J. (1966). *Toward a theory of instruction.* Cambridge, MA: Harvard University Press.

Carey, S. (1986). The acquisition of scientific knowledge--the problem of reorganization. In S. Strauss (Ed.), *Ontogeny, phylogeny, and the history of science.* Norwood, NJ: Ablex.

Chi, M. T. H. (1992). Conceptual change within and across ontological categories: Implications for learning and discovery in science. In R. N. Giere (Ed.), *Cognitive models of science: Minnesota studies in the philosophy of science (vol. 15).* Minneapolis, MN: University of Minnesota Press.

Chi, M. T. H. (2000). Misunderstanding emergent processes as causal, *Paper presented at the Annual Conference of the American Educational Research Association.*

Clough, E. E., & Driver, R. (1985). Secondary students conceptions of the conduction of heat: Bridging together scientific and person views. *Physics Education, 20*(4), 176-182.

Cotignola, M. I., Bordogna, C., Punte, G., & Cappannini, O. M. (2002). Difficulties in learning thermodynamic concepts: Are they linked to the historical development of this field? *Science & Education, 11*, 279-291.

Creswell, J. W., & Clark, V. L. P. (2007). *Designing and conducting mixed methods research.* Thousand Oaks, CA: Sage.

Donovan, M. S., & Bransford, J. D. (Eds.). (2005). *How Students learn: Mathematics in the classroom*. Washington, DC: The National Academies Press.

Driver, R., & Russell, J. (1982). *An investigation in the idea of heat, temperature and change of state, of children between 8 and 14 years*. University of Leeds, Leeds.

Erickson, G., & Tiberghien, A. (1985). Heat and temperature. In R. Driver, E. Guesne & A. Tiberghien (Eds.), *Children's ideas in science* (pp. 52-83). Philadelphia, PA: Open University Press.

Erickson, G. L. (1979). Children's conceptions of heat and temperature. *Science Education, 63*, 221.

Erickson, G. L. (1980). Children's view point of heat: A second look. *Science Education, 64*, 323-336.

Gotignola, M., Bordogna, C., Punte, G., & Gappannini, O. M. (2002). Difficulties in learning thermodynamic concepts: Are they linked to the historical development of this field? *Science & Education, 11*, 279-291.

Gunstone, R. F., & Mitchell, I. J. (1997). Metacognition and conceptual change. In J. Mintzes, J. Wandersee & J. Novak (Eds.), *Teaching science for understanding* (pp. 133-163). San Diego: Academic Press.

Harrison, A. G., Grayson, D. J., & Treagust, D. F. (1999). Investigating a grade 11 student's evolving conceptions of heat and temperature. *Journal of Research in Science Teaching, 36*(1), 55-87.

Hewson, P., & Hewson, M. (1992). The status of students' conceptions. In R. Duit, F. Goldberg & H. Niedderer (Eds.), *Research in physics learning: Theoretical issues and empirical studies* (pp. 59-73). Kiel, Germany: Institute for Science Education.

Jones, M. G., Carter, G., & Rua, M. J. (2000). Exploring the development of conceptual ecologies: Communities of concepts related to convection and heat. *Journal of Research in Science Teaching, 37*(2), 139-159.

Kesidou, S., & Duit, R. (1993). Students' conceptions of the second law of thermodynamics: An interpretive study. *Journal of Research in Science Teaching, 30*(1), 85-106.

Laburu, C. E., & Niaz, M. (2002). A lakatosian framework to analyze situations of cognitive conflict and controversy in student's understanding of heat energy and temperature. *Journal of Science Education and Technology, 11*(3), 211-219.

Leite, L. (1999). Heat and temperature: An analysis of how these concepts are dealt with in textbooks. *European Journal of Teacher Education, 22*(1), 75-88.

Lewis, E. L. (1996). Conceptual change among middle school students studying elementary thermodynamics. *Journal of Science Education and Technology, 5*(1), 3-31.

Lewis, E. L., & Linn, M. C. (1994). Heat energy and temperature concepts of adolescents, adults, and experts: Implications for curricular improvement. *Journal of Research in Science Teaching, 31*(6), 657-677.

Ling, M. L., Chik, P., & Pang, M. F. (2006). Patterns of variation in teaching the colour of light to primary 3 students. *Instructional Science, 34*, 1-19.

Lubben, F., Netshisaulu, T., & Campbell, B. (1999). Student's use of cultural metaphors and their scientific understandings related to heating. *Science Education, 87*, 761-774.

Mak, S. Y., & Young, K. (1987). Misconceptions in the teaching of heat. *School Science Review, March*, 464-470.

Niaz, M. (2006). Can the study of thermochemistry facilitate students' differentiation between heat energy and temperature? *Journal of Science Education and Technology, 15*(3), 269-276.

Romer, R. H. (2001). Heat is not a noun. *American Journal of Physics, 69*(2), 107-109.

Slone, M., Tredoux, C., & Bokhorst, F. (1996). Decalage effects for heating and cooling: A cross-cultural study. *Journal of Cross-Cultural Psychology, 27*(1), 51-66.

Stavy, R., & Berkovitz, B. (1980). Cognitive conflict a basis for teaching quantitative aspects of the concept of temperature. *Science Education, 64*, 679-692.

Summers, M. K. (1983). Teaching heat - an analysis of misconceptions. *School Science Review, June*, 670-675.

Tarsitani, C., & Vicentini, M. (1996). Scientific mental representations of thermodynamics. *Science and Education, 5*, 51-68.

Tashakkori, A., & Teddlie, C. (Eds.). (2003). *Handbook of mixed methods in social and behavioral research*. Thousand Oaks, CA: Sage.

Tiberghien, A. (1980). Modes and conditions of learning. An example: The learning of some aspects of the concepts of heat. In W. F. Archenhold, R. H. Driver, A. Orton & C. Wood-Robinson (Eds.), *Cognitive development research in science and mathematics* (pp. 288-309). Leeds, UK: University of Leeds Printing Service.

Tiberghien, A. (1983). Critical review of the research aimed at elucidating the sense that notions of temperature and heat have for students aged 10 to 16 years, *Proceedings of The first International Workshop Research on Physics Education* (pp. 73-90). Paris: Editions du CNRS.

Tiberghien, A. (1985). Heat and temperature. In R. Driver, E. Guesnes & A. Tiberghien (Eds.), *Childrens' ideas in science* (pp. 52-84). Milton Keynes: Open University Press.

Tomasini, G., & Balandi, P. (1987). Teaching strategies and children's science: An experiment on teaching "hot and cold", *Proceedings of the second international seminar "misconceptions and educational strategies in science and mathematics* (Vol. II, pp. 158-171). Ithaca, NY: Cornell University.

Veiga, M., Costa Pereira, D., & Maskill, R. (1989). Teachers' language and pupils' ideas in science lessons: Can teachers avoid reinforcing wrong ideas? *International Journal of Science Education, 11*(4), 465-479.

Wiser, M. (1986). *The differentiation of heat and temperature: An evaluation of the effect of microcomputer teaching on students' misconceptions*. Cambridge, MA: Harvard Graduate School of Education.

Wiser, M. (1988). The differentiation of heat and temperature: History of science and novice-expert shift. In S. Strauss (Ed.), *Ontogeny, phylogeny, and historical development* (pp. 28-48). Norwood, NJ: Ablex Publishing Corporation.

Wiser, M., & Amin, T. (2001). "is heat hot?" inducing conceptual change by integrating everyday and scientific perspectives on the thermal phenomena. *Learning and Instruction, 11*, 331-353.

Wiser, M., & Carey, S. (1983). When heat and temperature were one. In D. Gentner & A. Stevens (Eds.), *Mental models* (pp. 267 - 298). Hillsdale, NJ: Erlbaum.

APPENDIX I: LIST OF THE TEXTBOOKS IN THE STUDY

Der Karlsruher Physikkurs 1. Energie, Impuls, Entropie. Ein Lehrbuch für die Sekundarstufe 1. Aulis Verlag Deubner 2003.
Dorn.Bader Physik: Gymnasium SEK I. Hannover: Schroedel 2001.
Dorn.Bader Physik: Gymnasium SEK II. Hannover: Schroedel 2000.
Kuhn Physik Band 1.1. Braunschweig: Westermann Schulbuchverlag 1996.
Kuhn Physik Band II: Ergänzungsband Thermodynamik. Braunschweig: Westermann Schulbuchverlag 1993.
Kuhn Physik Band II 1. Teil: Klasse 11. Braunschweig: Westermann Schulbuchverlag1989.
Kuhn Physik Band II: Lehrerband. Braunschweig: Westermann Schulbuchverlag 1994.
Lehrbuch Physik: Sekundarstufe 2, Gesamtband. Cornelsen Volk und Wissen Verlag.
Metzler Physik (J. Grehn and J. Krause). Hannover: Schroedel 1998.
Physik für Gymnasien: Sekundarstufe I, Teilband 2. Berlin: Cornelsen Verlag 1994.

APPENDIX II: THE 2ND-PHASE QUESTIONNAIRE FOR GERMAN STUDENTS

Die folgende Fragen befassen sich mit den Konzepte und Phänomenen von Wärme. Bis auf Frage 9 gibt es jeweils vier verfügbare Antworten. Kreuzen Sie bitte die Aussage an, die Ihrer Meinung nach richtig ist. Die Antworten können eine, mehrere oder keine richtige Aussage Aussage enthalten. Falls Sie keine Aussage für richtig halten, wählen sie die Antwort „Sie sind alle falsch." und geben Sie ihre eigene Antwort.

Sie brauchen Ihren Namen nicht anzugeben; in jedem Fall bleiben die Ergebnisse der Befragung anonym und dienen nicht der Leistungsbewertung.

Name:_____ Jahrgang:_____ Geschlecht:_____
Falls Sie schon Grund- oder Leistungskurse gewählt haben: Ist Physik dabei? _____

Wir sind mit einer Anzahl von Phänomenen vertraut, die auf Wärmewirkung beruhen, wie das Kochen des Wassers oder die Ausdehnung der Körper bei Erwärmung. Was bedeutet Wärme für Sie?

☐ Wärme ist Entropie, das kann man mit Hilfe der Masse und Temperatur ausrechnen.

☐ Wärme ist die Bewegung der Teilchen des Materials und die daraus entstehende Energie.

☐ Wärme ist Hitze. Gegenteil von Kälte.

☐ Wärme bedeutet im Allgemeinen den Anstieg der Temperatur bzw. die Ausdehnung eines Körpers.

☐ Sie sind alle falsch. Meine Aussage:

_____

Unter welchen Bedingungen wird Wärme von einem Gegenstand auf einen anderen übertragen, wenn die Gegenstände sich berühren?

☐ Beide Gegenstände müssen dazu in der Lage sein, die Wärme zu transportieren (z.B. Elektronen im Metallgitter).

☐ Wenn die Gegenstände unterschiedlich große Wärmeenergie besitzen. Übertragung von Wärme bedeutet Energieübertragung des wärmeren Körper (höhere Energie) an den kälteren Körper (geringere Energie).

☐ Die Wärme wird übertragen, wenn ein Körper kälter ist als ein anderer. Dann wird die Wärme von dem wärmeren Körper abgeben, bis beide Gegenstände die selbe Temperatur haben.

☐ Einer der Gegenstände muss Wärme besitzen, also warm oder heiß sein, damit Wärme übertragen werden kann .

☐ Sie sind alle falsch. Meine Aussage:

_____

Wir verwenden bei Experimenten häufig Thermometer. Was ist die Funktion eines Thermometers?

☐ Das Thermometer kann die Temperatur eines Gegenstands und ihre Änderung messen.

☐ Ein Thermometer kann Wärme bzw. Kälte messen.

Appendix II 83

▫ Das Thermometer kann anzeigen, wie viel Wärme ein Körper hat. Wenn zum Beispiel in einem Glas 20g Wasser sind und wir lesen 20°C ab, so können wir schließen, dass das Wasser 400 Cal besitzt.

▫ Wenn wir mit dem Thermometer zwei verschiedene Gegenstände A und B messen, und das Ergebnis zeigt, dass Gegenstand A eine höhere Temperatur hat als Gegenstand B, dann können wir daraus schließen, dass A mehr Wärme besitzt als B.

▫ Sie sind alle falsch. Meine Aussage:

_____

_____

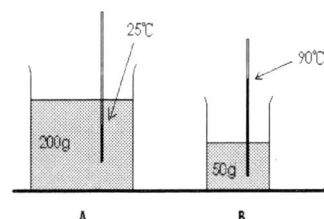

Wie die Abbildung zeigt, haben wir zwei Gläser mit (A) 200g Wasser bei 25°C und (B) 50g Wasser bei 90°C. Was kann man mit diesen Informationen über die Wärme sagen?

▫ Glas A mit 5000 Cal besitzt mehr Wärme als Glas B mit 4500 Cal.

▫ In Bezug auf das Volumen der beiden Stoffe hat B mehr Wärme (A: 25/200 < B: 90/50=360/200).

▫ Wenn wir diese zwei Gläser Wasser zusammen schütten, wird sich eine komplett neue Temperatur zwischen 25°C und 90°C ergeben. Das heißere Wasser kühlt ein wenig ab, und das kühlere erwärmt sich.

▫ Wenn wir diese zwei Gläser mit Wasser zusammen schütten, gleicht die Temperatur sich aus, und die Wärmeenergien addieren sich.

▫ Sie sind alle falsch. Meine Aussage::

_____

_____

Warum fühlt sich Metall allgemein kälter an als Holz?

▢ Weil Holz die Wärme speichert, bzw. die Wärmeaufnahme von Holz größer ist.

▢ Weil Metall eine höhere Dichte hat. Es ist fester, daher nimmt es nicht so viel Wärme auf und ist kälter.

▢ Weil Metall ein Leiter ist und Holz nicht. Metall nimmt Kälte auf.

▢ Weil Metall aus Teilchen besteht, die überwiegend sehr empfindlich sind und eher kalt sind.

▢ Sie sind alle falsch. Meine Aussage:

_____

_____

Wenn wir ein Ende eines Metallstabs erhitzen (siehe Abbildung), wird das andere Ende bald auch heiß. Was geschieht mit den Teilchen des Stabes während dieser Zeit?

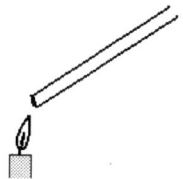

▢ Die Teilchen geraten in Bewegung und dehnen sich deshalb aus und wandern zum anderen Ende des Metallstabs.

▢ Die Teilchen nehmen Wärme auf und geben diese an ihre Nachbarteilchen teilweise wieder ab.

▢ Die Teilchen werden in Bewegung gesetzt und reiben sich aneinander. Dabei entsteht Wärme.

▢ Die Wärme nimmt immer mehr vom Volumen des Stabes ein. Sie wird die Kälte soweit ausgleichen, bis der Stab komplett warm ist. Die Teilchen erhitzen sich ebenfalls.

☐ Sie sind alle falsch. Meine Aussage:

_____
_____

Nehmen Sie an, dass es Sommer ist. Wir stellen einen Metallstab und einen Holzstab mit Raumtemperatur (ca. 20°C) eine Zeitlang nach draußen in die Mittagssonne (ca. 35°C); Was geschieht dann mit den Stäben?

☐ Die Temperatur des Metall- und des Holzstabes nähert sich 35°C an.

☐ Der Metallstab nimmt Wärme auf und die Temperatur steigt. Der Holzstab verändert sich dagegen nicht oder sehr wenig.

☐ Der Metallstab erhitzt sich schneller und besitzt deshalb nach einer Weile mehr Wärmeenergie als der Holzstab.

☐ Der Metallstab erhitzt sich schneller als der Holzstab, kann aber die Wärme nicht so lange speichern.

☐ Sie sind alle falsch. Meine Aussage:

_____
_____

Nehmen Sie an, dass wir eine Metallscheibe, eine Schüssel Mehl und eine Tasse Wasser bei Raumtemperatur haben (ca. 20°C). Wir stellen sie jetzt in einen Gefrierschrank (Innentemperatur: ca.-5°C). Was wird mit dem Metall, dem Mehl und dem Wasser geschehen?

☐ Alle Gegenstände werden sich der Temperatur des Gefrierschranks (-5°C) annähern.

☐ Alles wird sich abkühlen. Das Wasser gefriert. Das Metall wird extrem kalt werden. Das Mehl behält seine Temperatur, höchstens leicht kalt.

☐ Alle Gegenstände geben Wärme ab. Am Ende hat Metall die geringste Wärme, und das Mehl die höchste.

☐ Alle Gegenstände nehmen Kälte auf. Metall kann es am besten, das Wasser am zweitbesten und das Mehl am schlechtesten.

☐ Sie sind alle falsch. Meine Aussage:

_____
_____

Wenn wir das Eis aus dem Gefrierschrank nehmen, fängt es sofort an zu schmelzen. Erklären Sie (1) warum, (2) was ist der Unterschied zwischen den Teilchen des noch festen Eis und des schon geschmolzenen Eises? (Bitte schreiben Sie auf der Rückseite weiter.)

INDEX

A

absolute zero, 25, 61
absorption, 38
accuracy, viii, 13, 16
adolescents, 77
adults, 77
age, 10
air, 3, 4, 5, 20, 26, 40
alternative, 3, 7, 8, 21, 39
American Educational Research Association, 75
amplitude, 58, 63
anther, 59
appendix, 46
application, 1, 17, 57, 59
argument, 29, 39, 47, 53, 66
atoms, 40
attention, viii, 2, 45, 51, 54, 56, 61, 66, 74

B

bank account, 53
behavior, 27, 40, 57
bonding, 42, 58
bonds, 42

C

caloric theory, 65, 66, 72
carrier, 26
children, 4, 29, 75, 76, 78
Chinese, 22, 66
classes, 8
classroom, vii, 1, 8, 74, 76
codes, 46
cognitive, 76
coherence, viii, 5, 74
collisions, 60
commercial, 46
community, 65
competence, 1
conception, 21
conceptualization, 15, 69
concrete, 24
conduction, 27, 42, 51, 60, 61, 62, 63, 75
conductivity, 15, 31, 42
conductor, 6, 31, 32
conflict, 61, 74, 76, 77
confusion, vii, 3, 5, 11, 15, 16, 22, 25, 28, 40, 42, 46, 48, 63, 67, 73, 74
Congress, iv
convection, 60, 61, 76
cooking, 52
cooling, 6, 48, 77
credibility, 10
critical temperature, 38

cross-cultural, 77
cultural, vii, 2, 74, 77
culture, 7

D

data analysis, 12
data collection, vii, 2, 13
decay, 72
definition, 8, 16
degree, 5, 15, 17, 18, 19, 20, 33, 34, 35, 53, 54, 56, 57, 72
delta, 23
density, 31, 32, 38, 42
differentiation, 63, 66, 69, 72, 77, 78
diversity, 16

E

Education, 75, 76, 77, 78
educational research, 1
educators, 8
electricity, 16, 30
electromagnetic, 60
electromagnetic wave, 60
electronic, iv
electrons, 60
electrostatic, iv
energy, vii, 7, 12, 15, 16, 17, 18, 19, 23, 24, 25, 26, 27, 29, 30, 37, 39, 40, 42, 46, 47, 48, 49, 50, 51, 52, 53, 54, 55, 57, 58, 59, 60, 61, 62, 65, 66, 67, 69, 70, 71, 74, 76, 77
energy transfer, 60
English, 10, 22
Enthalpy, 52
entropy, 17, 47, 50, 70
environment, 5, 10, 16, 19, 21, 34, 35, 36, 37, 41, 48, 52
equilibrium, 2, 5, 8, 15, 16, 24, 25, 33, 34, 35, 36, 37, 38, 41, 43, 47, 55, 56, 57, 69, 72
European, 76
evidence, 3, 5, 16, 29, 42, 53
evolution, 65
evolutionary, 6, 7

expert, iv, 78
experts, 1, 50, 77

F

factual knowledge, 1
false, 10, 12
family, 7
feelings, 54
fire, 15, 16, 17, 18, 23
flow, 8, 21, 26, 28, 41, 50, 53, 61, 70
fluid, 3, 4, 8, 17, 21, 39, 41, 61, 71
freezing, 54
friction, 16, 27, 40, 52, 63, 67, 68, 72

G

gas, 57, 61
gases, 58
gauge, 54
generation, 63
Germany, vii, 2, 10, 17, 45, 53, 76
glass, 11, 24, 25
groups, 13, 31, 33, 35
guidelines, 45

H

hands, 54, 73
hardness, 4, 6, 18
Harvard, 75, 78
heat, vii, 2, 3, 4, 5, 6, 7, 8, 10, 11, 12, 13, 15, 16, 17, 18, 19, 20, 21, 22, 23, 24, 25, 26, 27, 28, 29, 30, 31, 32, 33, 34, 35, 36, 37, 38, 39, 40, 41, 42, 45, 46, 47, 48, 49, 50, 51, 52, 53, 55, 58, 59, 60, 61, 62, 63, 65, 66, 67, 68, 69, 70, 71, 72, 73, 74, 75, 76, 77, 78
heat conductivity, 42
heat release, 59
heat transfer, vii, 2, 5, 6, 11, 12, 13, 15, 16, 21, 25, 26, 27, 29, 30, 33, 41, 42, 48, 49, 59, 60, 61, 63, 68, 69, 70, 72, 73

heating, 6, 8, 15, 17, 21, 26, 48, 50, 53, 63, 67, 68, 77
high school, 10
hot water, 7, 54, 73

I

ice, 5, 12, 17, 18, 20, 22, 23, 26, 36, 37, 38, 39, 40, 41, 42, 72
id, 33
in transition, 23, 47, 48, 71
infinite, 26
injury, iv
instruction, 2, 3, 5, 8, 19, 74, 75
instructional design, vii, 2
instruments, 9
intensity, 23, 61, 70
interaction, 8
interactions, 6
intermolecular, 67
international, 78
interrelationships, 13
interviews, 5
intuition, 25

J

judge, 54

K

Keynes, 78
kinetic energy, 39, 42, 48, 52, 57, 62, 67

L

language, 8, 43, 50, 51, 52, 74, 78
law, 76
lead, 3, 24, 46, 56, 59, 66, 67
learners, 66
learning, vii, viii, 1, 2, 7, 8, 45, 57, 59, 66, 70, 72, 74, 75, 76, 77
learning process, 1, 2

linkage, 15, 17, 29, 38, 39, 45, 56, 66
liquid phase, 13
liquids, 58, 70

M

magnetic, iv
mathematical, 11, 16, 49, 59
mathematics, 77, 78
meanings, viii, 2, 7, 11, 15, 16, 23, 45, 46, 49, 50, 51, 52, 54, 57, 67, 71, 74
measures, 4, 5
mechanical, iv, 53, 67
mechanical energy, 53, 67
mechanics, 67
melt, 18, 23, 36, 38, 40, 41, 72
melting, 12, 17, 18, 20, 22, 23, 26, 36, 37, 38, 39, 40, 41, 42, 58
melts, 5, 19, 21, 39, 40
mental representation, 77
metals, 5
metaphor, 53
metaphors, 53, 77
Minnesota, 75
misconception, 5, 20, 25, 48, 50, 56, 63, 69
misconceptions, vii, 2, 5, 7, 8, 32, 37, 45, 46, 52, 59, 66, 67, 74, 77, 78
misleading, 48, 52
misunderstanding, 66
mixing, 24, 59, 73
models, 75, 78
molecules, 26, 57
money, 53
motion, 16, 21, 23, 27, 38, 39, 40, 41, 42, 57, 71
movement, 4, 6, 21, 22, 26, 39, 53, 57, 61, 66
MPI, 46, 53, 55, 56, 61
multiple-choice questions, 10, 11, 12, 13

N

natural, 5, 6, 39, 59
New York, iii, iv
Newtonian, 67

NPI, 46, 54, 60, 65, 66

O

observations, 73

P

Paper, 75
parallelism, 39, 74
Paris, 77
particles, 11, 12, 13, 16, 18, 19, 20, 22, 26, 27, 28, 32, 36, 37, 38, 39, 40, 41, 42, 52, 53, 57, 58, 60, 62, 63, 66, 67, 70
perception, 54
personal, 10, 36
personal accounts, 10, 36
Philadelphia, 76
philosophical, 9
philosophy, 75
phylogeny, 75, 78
physics, 2, 67, 76
PII, 62
plastic, 5, 7
play, 3
poor, viii, 6, 16
population, 11
potential energy, 48, 53, 58, 67
pragmatic, 9
pragmatism, 9
prediction, 25
preparation, iv
property, iv, 4, 5, 7, 30, 53, 55, 56, 70
publishers, 51

Q

quantitative research, 9
questionnaire, 9, 10, 11, 12, 16

R

radiation, 12, 27, 51, 60, 61, 63, 67, 70

range, 54, 57, 72, 73
reality, 70
recall, 26
reconciliation, 16, 28
reconstruction, vii, 70, 71
reflection, 40
research, vii, 2, 3, 5, 7, 9, 53, 66, 75, 77
researchers, 1, 9
resources, 7
response format, 10
revolutionary, 7
room temperature, 21, 32, 33, 38, 40

S

sample, 17, 19, 24, 30, 31, 33, 35, 38, 39, 42, 43, 45, 47, 69, 74
satellite, 10
school, vii, 2, 5, 8, 10, 42, 77
science, vii, viii, 2, 5, 8, 11, 15, 17, 19, 25, 36, 45, 50, 59, 66, 69, 71, 75, 76, 77, 78
science educators, 8
scientific, vii, 2, 6, 8, 32, 42, 45, 46, 47, 50, 51, 54, 65, 66, 67, 71, 73, 74, 75, 77, 78
scientific community, 65
scientific knowledge, vii, 2, 71, 75
scientific understanding, 77
scientists, 5, 6, 65, 66
secondary students, vii, 2, 6, 8, 10, 15, 29, 36, 69
sensation, 54
series, 10
services, iv
shape, 1, 24, 29, 56
shares, 62, 74
similarity, 22
social, 77
solid phase, 41
specific heat, 15, 30, 31, 32
speed, 4, 6, 20, 31, 42, 55, 57, 58, 61, 62
SPSS, 13
storage, 48, 71
strategies, 78
strength, 1, 36
stress, 60

structuring, 53
student group, 11, 45
students, vii, 1, 2, 3, 5, 6, 7, 8, 9, 10, 11, 12, 15, 16, 17, 18, 19, 20, 22, 23, 24, 25, 26, 27, 28, 29, 30, 32, 33, 34, 35, 36, 37, 38, 39, 40, 41, 42, 43, 45, 46, 48, 49, 50, 52, 53, 56, 57, 59, 61, 63, 66, 67, 69, 70, 71, 72, 73, 74, 75, 76, 77, 78
students' ideas, vii, 8, 45
students' understanding, vii, 2, 8, 11, 12, 25, 33, 36, 50, 52, 57, 63, 69, 73
subjective, 9
substances, 4, 5, 6, 25
sugar, 5
summaries, 52
summer, 52
sunlight, 55
supply, 55
surface area, 5
systems, 4, 16, 28, 29, 35

thermal energy, 15, 22, 50, 52, 53, 66, 67, 70, 74
thermal equilibrium, 2, 5, 8, 15, 16, 24, 25, 33, 35, 38, 41, 43, 47, 55, 56, 57, 69, 72
thermodynamic, 75, 76
thermodynamics, vii, 2, 3, 7, 54, 57, 76, 77
thinking, 48
time, 6, 11, 47, 49, 53, 56, 66
tolerance, 54, 73
transfer, vii, 2, 5, 6, 10, 11, 12, 13, 15, 16, 21, 25, 26, 27, 28, 29, 30, 33, 35, 41, 42, 47, 48, 49, 53, 59, 60, 61, 63, 67, 68, 69, 70, 72, 73
transition, 11, 12, 13, 16, 23, 32, 36, 37, 38, 47, 48, 58, 71, 72
transitions, 65
transmission, 26
transport, 61
trend, 10
trust, 25

T

Taiwan, vii, 2, 10, 45, 47
Taiwanese students, 11, 12, 17, 19, 21, 24, 27, 29, 30, 31, 32, 33, 34, 36, 38
teachers, 2, 8, 45, 48, 51, 56, 66, 78
teaching, vii, viii, 2, 6, 7, 8, 45, 57, 60, 70, 74, 77, 78
temperature, vii, 2, 3, 5, 6, 7, 8, 10, 11, 12, 15, 16, 17, 18, 19, 20, 22, 23, 24, 25, 26, 29, 30, 31, 32, 33, 34, 35, 36, 37, 38, 39, 40, 41, 42, 43, 46, 47, 48, 50, 51, 52, 53, 54, 55, 56, 57, 58, 59, 60, 61, 62, 63, 66, 67, 69, 70, 71, 72, 73, 75, 76, 77, 78
textbooks, vii, 2, 7, 43, 45, 46, 47, 48, 49, 50, 52, 53, 54, 55, 56, 57, 58, 59, 60, 61, 63, 65, 66, 67, 69, 70, 76
theory, 1, 6, 8, 13, 48, 65, 66, 67, 71, 75
thermal, vii, 2, 3, 5, 6, 8, 10, 11, 12, 15, 16, 17, 19, 22, 23, 24, 25, 33, 35, 37, 38, 41, 43, 45, 46, 47, 50, 52, 53, 55, 56, 57, 59, 66, 67, 69, 70, 72, 73, 74, 78

U

UK, 77

V

values, 54
variables, 13
variation, 77
vibration, 59, 63
visible, 71, 72

W

Washington, 76
water, 11, 18, 20, 21, 22, 23, 24, 25, 26, 33, 34, 36, 37, 38, 39, 40, 41, 42, 47, 52, 54, 55, 59, 73
weakness, 60, 73
wood, 5, 13, 30, 31, 32, 33, 34, 35, 38, 72
worldview, 9